EXTINCTION STUDIES

COLUMBIA UNIVERSITY PRESS   *New York*

EXTINCTION STUDIES

Stories of Time,
Death, and Generations

EDITED BY
Deborah Bird Rose,
Thom van Dooren,
and Matthew Chrulew

COLUMBIA UNIVERSITY PRESS
Publishers Since 1893
New York   Chichester, West Sussex
cup.columbia.edu

Library of Congress Cataloging-in-Publication Data

Names: Rose, Deborah Bird, 1946– editor. | Van Dooren, Thom, 1980–
    editor. | Chrulew, Matthew, editor.
Title: Extinction studies : stories of time, death, and generations / edited by
    Deborah Bird Rose, Thom van Dooren, and Matthew Chrulew.
Description: New York : Columbia University Press, 2017. | Includes
    bibliographical references and index.
Identifi ers: LCCN 2016050679 | ISBN 9780231178808 (cloth) |
    ISBN 9780231178815 (pbk.) | ISBN 9780231544542 (e-book) Subjects:
LCSH: Extinction (Biology) | Endangered species. | Wildlife
    conservation.
Classification: LCC QH78 .E96 2017 | DDC 576.8/4—dc23
LC record available at https://lccn.loc.gov/2016050679

COVER IMAGE: © ISABELLA KIRKLAND, *GONE*, 2004. FROM THE ARTIST'S
TAXA SERIES
BOOK & COVER DESIGN: CHANG JAE LEE

CONTENTS

FOREWORD
Cary Wolfe   vii

Introduction: Telling Extinction Stories
Deborah Bird Rose, Thom van Dooren, and Matthew Chrulew   1

1. Walking with Ōkami, the Large-Mouthed Pure God
   James Hatley   19

2. Saving the Golden Lion Tamarin
   Matthew Chrulew   49

3. Extinction in a Distant Land:
   The Question of Elliot's Bird of Paradise
   Rick De Vos   89

4. Monk Seals at the Edge:
   Blessings in a Time of Peril
   Deborah Bird Rose   117

5. Encountering Leatherbacks in Multispecies Knots
   of Time
   Michelle Bastian   149

6. Spectral Crows in Hawai'i:
   Conservation and the Work of Inheritance
   Thom van Dooren   187

   Afterword: It Is an Entire World That Has Disappeared
   Vinciane Despret   217

   CONTRIBUTORS   223
   INDEX   227

FOREWORD

CARY WOLFE

What kind of event is extinction? To address this question, we have to begin with an assertion exemplified, in different ways, in the book you are about to read, one that will seem paradoxical to some and commonsensical to others: that extinction both is the most natural thing in the world and, at the same time, is not and never could be natural. On the one hand, as noted more than once in these pages, 99.9 percent of all species that have ever existed in the history of this planet are extinct; on the other hand, extinction can hardly be regarded as "natural" in any simple sense, and not just because, as any number of people have argued, "nature," conceived as a realm apart, untouched and unshaped by human affairs, ceased to exist a long time ago, as all the recent talk about climate change and the Anthropocene makes clear.[1] Beyond this, the psychoanalytically inclined among us point out that any human registration of the so-called fact of nature is always already radically denaturalized because the symbolic and imaginary realms that make the presence of nature manifest to us in their different ways are anything but "natural."[2]

This is precisely the point—or one of the points—of the more anthropological and ethnographic orientation in this volume toward "extinction *stories*," which remind us, the editors write, that "there is no singular phenomenon of extinction; rather, extinction is experienced, resisted, measured, enunciated, performed, and narrated in a variety of ways to which we must attend." In short, extinction—whatever else it may be—is *never* a generic event and is *always* a multi-contextual phenomenon requiring multi-disciplinary modes of encounter and understanding. That fact is worth remembering when we ask the question: When a being, human or non-human, dies, what goes out of the world? What is lost to the world? And what world are we left with? As James Hatley notes in his chapter on the Honshu wolf of Japan, the Ōkami, "Something like Ōkami should not disappear in our time without its departure having been noticed, without its loss mattering. . . . A question of etiquette is involved, 'transhuman etiquette.'" One of our best exemplars of "transhuman etiquette," Vinciane Despret, provides her own eloquent and moving answer to these questions in the afterword to this collection. Meditating on the extinction of the Passenger Pigeon. She writes:

> The world dies from each absence; the world bursts from absence. For the universe, as the great and good philosophers have said, the entire universe thinks and feels itself, and each being matters in the fabric of its sensations. Every sensation of every being of the world is a mode through which the world lives and feels itself, and through which it exists. And every sensation of every being of the world causes all the beings of the world to feel and think themselves differently. When a being is no more, the world narrows all of a sudden, and a part of reality collapses. Each time an existence disappears, it is a piece of the universe of sensations that fades away.

Among contemporary philosophers, no one has meditated on this question more, perhaps, than Jacques Derrida (2005), and in response he has offered a rather different—indeed, seemingly counter-intuitive—assertion:

> [D]eath is nothing less than an end of *the* world. Not *only one* end among others, the end of someone or something *in the world*, the end of a life or a living being. . . . Death marks each time, each time in defiance of arithmetic, the absolute end of the one and only world, of that which each opens as a one and only world, the end of the unique world, the end of the totality of what is or can be presented as the origin of the world for any unique living being, be it human or not. (140)

Now such an assertion may seem at first glance to be an excessively Heideggerian hangover on Derrida's part, but from another vantage point, Derrida (1986) may be seen here as trying to move us away from what he calls the "dogma" (173) of Heidegger's famous (or infamous) investigations of the differences among humans, animals, and stones with regard to the question of "world," and toward the necessity of tending to different ways of being in the world for different creatures.

No one limns Heidegger's attitude toward the difference between scientific and philosophical knowledge better than Derrida (2011) himself in the seventh session of the second year of his seminars in *The Beast and the Sovereign*, where he notes that Heidegger's "strange concept" of the animal's "*poverty in world* . . . does not consist in a quantitative relation of degree, of more or less" (192). "About this presupposed essence," he continues, Heidegger presumes that "the zoologist, the zoologist *as such* at least, has nothing to say to us" (194). In fact, as Michael Naas (2015) has pointed out, Derrida is often quick to note that Heidegger, more than most philosophers, takes "into account a certain ethological knowledge" with regard to animals (149). But from Derrida's point of view, that only makes all the more dogmatic Heidegger's "thesis of essence ('the animal is poor in world') independent of zoological knowledge," a thesis that pertains, as Derrida (2011) puts it, to "the *animal in general*," to "*every* animal" as "*equally* poor in world" (196–197). Instead, Derrida's (2008) focus is not on "effacing the limit" between human and nonhuman animals (or, indeed, between different forms of life, human and nonhuman), but rather "in multiplying its figures, in complicating, thickening, delinearizing, folding, and dividing the line precisely by making it increase and multiply" (28; see also 34–35), eventuating in the sorts of "knots" and "interfaces" between different living beings, across the generations, invoked in the stories told in these pages.

And Derrida's observation draws our attention to something more, something easier to overlook, with its passing phrase, "each time in defiance of arithmetic": *how* we think, name, and categorize non-human life matters, and not just as an element in an academic exercise. As Hatley observes,

> The very notion that one can put one's finger upon a name on a list and decide, just because, that this is the animal one will think upon, seems itself indicative of the plight of animals in the Anthropocene—to be surrounded by human beings for whom the perplexity and complexity of the living world has been reduced to an amorphous set of words and a collection of fleeting images. The very practices by which the living world finds its place in human thought is increasingly dominated by a false familiarity in which everything is brought near to the human thinker precisely by its having been first stripped of its manifold living senses and so reduced to a bare minimum of meaning.

These sorts of knowledge-making practices not only take us out of the world—paradoxically, by seeming to make the entirety of the world "present at hand," as Heidegger put it—but also have real consequences for nonhuman (and human) beings. Lives are at stake. As Rick De Vos notes in his chapter on birds of paradise,

> The conceptualization and enunciation of species and hybridity work together so as to determine that the lives of some birds are more significant and more valuable than the lives of others and, by extension, the absence of their lives and ways of living. In the case of birds of paradise in the late nineteenth and early twentieth centuries, this enunciation helps to exculpate the perpetrators of the widespread massacre, downplaying the disappearance of particular birds as the loss of hybrids rather than the extinctions of species.

These points are worth remembering in the context of our current fascination with de-extinction projects involving the woolly mammoth, the Tasmanian tiger, the auroch, and, yes, the Passenger Pigeon. Here, the tendency has been to focus the question of extinction and de-extinction on the brute material presence or absence of the animal's body and genetic material, when in fact, from the more supple and complex point of view that animates this collection, a species may *already* be said to be extinct even though the last survivors of its kind live on. As Deborah Bird Rose (2013) has written, such survivors may be encountered in the "deathzone: the place where the living and the dying encounter each other in the presence of that which cannot be averted. Death is imminent but has not yet arrived" (3–4). Indeed, as Despret suggests at the opening of her coda to this volume, the Passenger Pigeon was in a very real sense *already* extinct before September 1, 1914, even though the last survivors of the species—Martha and her mate, George—persisted in their cage in the Cincinnati Zoo for a while longer.

From the vantage point of Michelle Bastian—and of Deborah Bird Rose and Thom van Dooren in foundational work undertaken previously—what this reveals is the far-from-generic quality of temporality itself, how textured and interwoven it is in a multi-species world where, "through sequences of generational time," as Bastian puts it, neither sequence nor synchrony "happens automatically, but both are embodied achievements" in the exquisitely co-ordinated and highly punctuated relations that "bring together food and fed, pollinator and pollinated, traveler and medium traveled." What Rose calls these "multispecies knots of time" are part of what is endangered in an era of mass extinction, as the very fabric of time itself on Earth, woven over thousands and thousands of years, threatens to unravel before our very eyes—a phenomenon onto which we slap the label "Anthropocene."

In his book *Basin and Range*, John McPhee (1980) observes that geologists now "see the unbelievable swiftness with which one evolving species on earth has learned to reach into the dirt of some tropical island and fling 747s into the sky. . . . Seeing a race unaware of its own instantaneousness in time, they can reel off all the species that have come and gone, with emphasis on those that have specialized themselves to death" (133–134). In this light, if our current modes of technoscience and instrumental reason constitute, as Richard Beardsworth (1996) has argued, a historically unprecedented *acceleration* of time that risks normalizing "an experience of time that *forgets* time" (148), then tending to the qualitative difference of time on Earth—the fact that human time is simply one mode of temporality among many—involves calibrating ourselves to what we might call the different *speeds* of different forms of life: the slowness of the leatherback turtle, the speed of the viral vector or the bacterial network, the immense sonic architectures by which humpback whales reconfigure time and space on a literally global scale. Indeed, one of the great virtues of the explosion of interest in epigenetics and the microbiome is to remind us that those differential speeds obtain and express themselves *in us*, in ways that we are only beginning to understand. We are "knots of time," which our dominant modes of experiencing time encourage us to forget.

But what if the Anthropocene—"that loaded term for the end to the dream/nightmare of a hyper-separated nature," as the editors call it—is just another form of hubris? What if "what calls *itself* man," as Derrida (2008:30) puts it, is nothing but the upshot of what Alan Stoekl (2007) calls "peak oil" (132)? What if we recognize, as he writes, that "the illusion we call 'Man' derives his 'freedom' from the quantification and commodification of natural resources: oil, to be sure, but also the steel, plastics and other materials that go to make up the 'autonomist' lifestyle" (Stoekl 2007:132)? And what is peak oil if not, precisely, a relation to time? What else but a hypercondensed form of that very slowness of nonhuman animals I mentioned a moment ago, but also of the ossified time (following the classical definition) of plants, which are differentiated from animals by means of their *immobility* (and thus are even slower than the animals themselves)? From that vantage point, what we call "human" is nothing other than the spectacular conflagration, the wanton burning, of time itself: not *our* time, because time is not *for* the human, but other times forcibly *made* our time, millions and millions of slow inhuman years released in a geological blink of an eye, the "luminous, explosive characteristics" of "humanity's presence on earth," as McPhee (1980) puts it, consisting now "not merely of the burst of population in the twentieth century, but of the whole millennial moment of people on earth—a single detonation, resembling nothing so much as a nuclear implosion" (132).

I have lingered on this tangle of questions to underscore what I take to be a point of *biopolitical* emphasis that I share with the editors of this collection—a point that often gets lost in discussions of the current mass-extinction event, which tend to present humans, as the editors observe, in a far too one-dimensional way, as "an amorphous and monotonal 'threat.'" But "to say that this is an *anthropogenic* extinction event," they continue, "is to pose a question: Which forms of human life are driving these catastrophic processes of loss, and in what other diverse ways are humans drawn into and implicated in extinction and its resistance?" What is needed is acknowledgment of "the specific political, economic, and cultural forms of human organization most responsible for any given extinction. In both contexts, radical inequity and highly differential positioning are the name of the game. Excavating this specificity matters." And in the biopolitical context of "making live" and "letting die" (to use Michel Foucault's [2008:271–272] famous formulation) that characterizes the current mass-extinction event and our responses to it in everything from conservation biology to de-extinction, these questions extend not just to what we do *to* non-human beings, but also to what we do *for* them. For as Matthew Chrulew observes of the vexed, often heartbreaking history of conservation efforts on behalf of the golden lion tamarin,

> Without denying the pressing need to resist the horror of extinction—indeed, in service to this very ambition—we must ask *at what price* (and courting what dangers, tolerating what failures) are species rescued from this fate? What costs were borne by these tamarins? Of what were they (made) capable—and incapable? Can we tell the story in a way that attends more widely to the multiplicity of these costs and achievements, and that problematizes them in turn?

Paying attention to these details, these "specificities," doesn't obviate John McPhee's (1980) sage observation:

> For establishing our bearings through time, we obviously owe an incalculable debt to vanished and endangered species, and if the condor, the kit fox, the human being, the black-footed ferret, and the three-toed sloth are at the head of the line to go

next, there is less cause for dismay than for placid acceptance of the march of prodigious tradition. The opossum may be Cretaceous, certain clams Devonian, and oysters Triassic, but for each and every oyster in the sea, it seems, there is a species gone forever. Be a possum is the message, and you may outlive God. (125–126)

But in the meantime, in the "knots of time" that braid us together with the myriad forms of life with which we share the planet at this very moment, the essays collected here ask: What is to be done? *For* whom? *By* whom? And at what cost?

NOTES

1. For one version of this argument, see Morton (2013).

2. As Slavoj Žižek (1992) put it, now nearly twenty-five years ago, "the fact that man is a speaking being means precisely that he is, so to speak, constitutively 'derailed,'" an "open wound of the world," as Hegel put it, that "excludes man forever from the circular movement of life," so that "all attempts to regain a new balance between man and nature" can only be a form of fetishistic disavowal (36–37).

REFERENCES

Beardsworth, Richard. 1996. *Derrida and the Political.* London: Routledge.

Derrida, Jacques. 1986. "*Geschlecht* II: Heidegger's Hand." Translated by John P. Leavey Jr. In *Deconstruction and Philosophy: The Texts of Jacques Derrida*, edited by John Sallis, 161–196. Chicago: University of Chicago Press.

——. 2005. "Rams: Uninterrupted Dialogue—Between Two Infinities, the Poem." In *Sovereignties in Question: The Poetics of Paul Celan*, edited by

Thomas Dutoit and Outi Pasanen, 135–163. New York: Fordham University Press.

——. 2008. *The Animal That Therefore I Am*. Edited by Marie-Louise Mallet. Translated by David Wills. New York: Fordham University Press.

——. 2011. *The Beast and the Sovereign*. Vol. 2. Edited by Michel Lisse, Marie-Louise Mallet, and Ginette Michaud. Translated by Geoffrey Bennington. Chicago: University of Chicago Press.

Foucault, Michel. 2008. *"Society Must Be Defended": Lectures at the Collége de France, 1978–1979*. Edited by Michel Senellart. Translated by Graham Burchell. New York: Palgrave Macmillan.

McPhee, John. 1980. *Basin and Range*. New York: Farrar, Straus and Giroux.

Morton, Timothy. 2013. *Hyperobjects: Philosophy and Ecology After the End of the World*. Minneapolis: University of Minnesota Press.

Naas, Michael. 2015. *The End of the World and Other Teachable Moments: Jacques Derrida's Final Seminar*. New York: Fordham University Press.

Rose, Deborah Bird. 2013. "In the Shadow of All This Death." In *Animal Death*, edited by Jay Johnston and Fiona Probyn-Rapsey, 1–20. Sydney: Sydney University Press.

Stoekl, Alan. 2007. *Bataille's Peak: Energy, Religion, and Postsustainability*. Minneapolis: University of Minnesota Press.

Žižek, Slavoj. 1992. *Looking Awry: An Introduction to Jacques Lacan Through Popular Culture*. Cambridge, Mass.: MIT Press.

EXTINCTION STUDIES

INTRODUCTION

Telling Extinction Stories

DEBORAH BIRD ROSE, THOM VAN DOOREN,

AND MATTHEW CHRULEW

At this very moment, many of Earth's living kinds are slipping away; sometimes quietly, sometimes in bright bursts of controversy, chaos, and pain. As a growing cohort of biologists is telling us, we are either already within or well on our way toward the sixth mass-extinction event since complex life evolved on this planet (Barnosky et al. 2011; Kingsford et al. 2009; Myers and Knoll 2001). While charismatic endangered species occasionally grab a headline or two, all around us a quieter systemic process of loss is relentlessly ticking on: hundreds, perhaps thousands, of species becoming extinct every year.

Mass extinction is marked by three primary characteristics: a radically high number of species being lost; the loss taking place across a diverse range of life-forms; and the compressed time frame within which it is occurring (Raup and Sepkoski 1982). This mass-death event differs from the previous five in one fundamental way: it is being driven almost entirely by humans, pressed along by relentless processes of habitat loss, direct exploitation, climate change, and more.

And yet, despite this central responsibility, people are involved in extinction in varied and ambivalent ways. We eat animals, log their forests for housing, cull their numbers for convenience, destroy and transform their homes and lives through unyielding systems of development and security. In this context, many people find themselves overwhelmed with the depressing inevitability and crushing finality of extinction. It is all the more astonishing, therefore, that along with sadness there is hope, along with seeming inevitability there is resistance. Scientists count creatures, tag them, relocate them to safer ground. Committed groups of all sorts, moved by the plight of a fellow being, work to protect the living and to slow the course of extinction. Many people query their own ethics and seek to live with less dire impact.

BIOCULTURAL RESPONSES

This book responds to the biocultural complexity of this time of extinctions. The chapters emerge out of the collaborative discussions of the Extinction Studies Working Group, a scholarly collective we formed around the shared conviction that our present time demands considered, lively, and creative responses from the humanities.[1] While extinction has been a topic of some, albeit limited, interest to scholars working in distinct disciplinary areas within the humanities—for example, environmental ethics, anthropology, literature, and history—our group has aimed to develop a distinctively *interdisciplinary* approach to this topic, grounded in the humanities but pushing out beyond them into a broader engagement with the social and natural sciences, as well as with the wider frames of understanding and meaning making that exist beyond the academy.

Through this work, we have sought to demarcate a general approach that we have called "extinction studies." This approach is grounded in the understanding that there is no singular phenomenon of extinction; rather, extinction is experienced, resisted, measured, enunciated, performed, and narrated in a variety of ways to which we must attend (De Vos 2007; van Dooren 2014). Our work, therefore, has centered on detailed case studies of complex processes of loss, exploring the "entangled significance" of extinction. As a result, this is not a book about extinction in the abstract; nor does it explore the complicity or responsibility of an amorphous "humanity." Rather, it is an effort to inhabit sites of incredible biological and cultural *diversity*—much of it, sadly, threatened—in a way that acknowledges that specificity matters. Each of the chapters draws on fieldwork, historical research, and/or cultural analysis, in combination with an engagement with biological and ecological literatures, to explore a particular species and its relationships with a larger, multispecies world. Through this situated approach, each chapter is an effort to tell a unique "extinction story," providing a narrative-based engagement that explores what an extinction means, why it matters, and to whom.

The stories presented here grapple with, and respond to, the complexity and ethical significance of specific sites of loss. As such, they do not aim to synthesize a universal picture or to reveal the experiences of either humans or nonhumans in any absolute sense. Rather, they aim to tell stories in ways that are open and accountable to these diverse others (Haraway 1991; van Dooren and Rose 2016). Our commitment to the storytelling mode is based on the fact that unlike many other modes of giving an account, stories can allow multiple meanings to travel alongside one another (Griffiths 2007). In addition, they can hold open possibilities and interpretations and refuse the kind of closure that prevents others from speaking or becoming (Smith 2005).[2] Of course, not all stories do this in practice, but the aim of this book is to tell stories that create openings, stories that can help us to inhabit *multiply*-storied worlds in a spirit of openness and accountability to otherness.[3]

As Donna Haraway and others tell us, storytelling is never innocent: it matters which stories we use to tell and think other stories with.[4] Telling multispecies stories of extinction is an inherently interdisciplinary task, one that draws us into conversation with a host of different ways of making sense of others' worlds. In large part, it is about, in Anna Tsing's (2011) terms, "passionate immersion in the lives of nonhumans" (29).[5] Our approach draws heavily on a subsection of the natural sciences within the fields of biology, ecology, and ethology, as well as emerging interdisciplinary work in "philosophical ethology" (Buchanan, Bussolini, and Chrulew 2014). In telling stories informed by these literatures, we invite readers into a sense of curiosity about the intimate particularities of others' ways of life: how they hunt or reproduce; how they relate to, craft, and make sense of their particular places; how they entice seed dispersers or navigate complex and often vast terrains.

Beyond the natural sciences, we also make use of the detailed observations and understandings of other knowledgeable peoples: from hunters and farmers, to artists, indigenous peoples, wildlife carers, and many others. In each of these cases, as with the insights of the natural sciences, knowledges are evaluated not only for what they teach us, but also for the particular political and technical architectures of framing within which they are produced. Our practice of moving beyond the academy and drawing on diverse perspectives is part of the critical work of decolonizing Western boundaries around knowledge and expertise (Apffel-Marglin and Marglin 1996).

Our effort to give "thick" accounts of other-than-human ways of life is in conversation with an extensive and growing body of work in the humanities and social sciences in the broad area of "multispecies studies."[6] This work is taking place under a range of labels, including "multispecies ethnography" (Kirksey and Helmreich 2010), "etho-ethnology" (Lestel, Brunois, and Gaunet 2006), "anthropology of life" (Kohn 2013), "anthropology beyond humanity" (Ingold 2013), "lively ethography" (van Dooren and

Rose 2016), and "more-than-human geographies" (Lorimer and Driessen 2014; Whatmore 2003). Despite their differences, all these approaches are united by a common interest in better understanding what is at stake—ethically, politically, epistemologically—for different forms of life caught up in diverse relationships of knowing and living together. "Extinction studies" is a kindred field of research, but one with a particular focus on understanding and responding to processes of collective death, where not just individual organisms, but entire ways and forms of life, are at stake.

While each of our extinction stories starts with a focus on a specific disappearing, or perhaps already disappeared, species, the openness of these accounts inevitably draws others into the frame. To this end, the chapters assembled here also draw heavily on ethnographic and historical methods to flesh out diverse people's understandings of and relationships with their changing world. Thus, in contrast to academic divisions that frequently split environmental problems into their "natural/technical" and "human/cultural" components, leaving only the latter for the social sciences (and, occasionally, the humanities), the extinction stories we tell here work across this entrenched and damaging divide. This means that our stories are attentive to the simultaneously biological and cultural complexity of our world, insisting that extinction is an inherently and inextricably *biocultural* phenomenon. Our modes of analysis must reflect this fact if they are to grapple with complex worlds-in-the-making (and unmaking) in meaningful ways.

As such, this is definitely *not* a project about the "human dimensions" of extinction. We do not thematize here the era of the "Anthropocene," that loaded term for the end to the dream/nightmare of a hyper-separated nature, though we recognize the concept's increasing diversification and intersection with questions of animal life (HARN Editorial Collective 2015). Nor do we concern ourselves with the projected extinction of humanity as a theoretical challenge to the excesses of humanism, as productive as such reflections might be (Colebrook 2014). Our focus is with non-human animals, whose extinctions are more than an admonishing fable, as much as we might learn from their shattering facticity (Chrulew and De Vos, forthcoming). Our wager is that engaging carefully and responsively with, in particular, these animal others—whose existence and disappearance, character and charisma, ways of living and being in the world have always been so central to the formation of our phylogenetic, social, and individual identity—will generate distinctive and meaningful ways not only of challenging human dominance, but of forming new modes of multispecies flourishing that engender hope and love in the face of such loss.

Nor does our project excise the humans. In this context, to say that this is an *anthropogenic* extinction event is to pose a question: Which forms of human life are driving these catastrophic processes of loss, and in what other diverse ways are humans drawn into and implicated in extinction—and its resistance? While we admire Elizabeth Kolbert's *The Sixth Extinction* (2014), and the important role that it has played in raising broader awareness of biodiversity loss, it is precisely these questions that her book fails to answer or even ask. Humans really appear in Kolbert's pages in only two forms: that of the (often heroic) conservationist struggling to hold onto disappearing species, and that of an amorphous and monotonal "threat." The many ways in which human communities are affected by and suffer through extinction are not present; nor is there a detailed discussion of the specific political, economic, and cultural forms of human organization most responsible for any given extinction. In both contexts, radical inequity and highly differential positioning are the name of the game. Excavating this specificity matters. As Susan Leigh Star (1990) would put it: Who wins and who loses—*cui bono?*—as our rich and biodiverse world is unraveled from the inside?

Ethical Responses

Along with offering analytic descriptions of complex entangle-ments, these stories aim to do important ethical and political work, exploring the particular constellations of life and death that are taking shape around the edges of extinction. From captive breeding facilities to protected areas, from wildlife carers to pil-grims and changing hunting regimes, much is at stake in these shifting spaces and practices, with their different forms of sacri-fice and varied measures of success. Each chapter attempts to navigate this complexity, to explore but also to critically analyze and ultimately to advocate, directly or through implication, for some possibilities and not others. Ultimately, this book asks how new accounts of extinction, accounts grounded in and attentive to biocultural complexity, might reanimate and reconfigure pos-sibilities for response and responsibility in this period of incred-ible loss.

It thus contributes to the recognition of the inadequacy outlined so elegantly by Cary Wolfe (2003): that the dominant "operative theories and procedures" we now have and through which we now articulate "the social and legal relation between ethics and action are inadequate . . . for thinking about the ethics of the question of the human as well as the nonhuman animal" (192–193). We are called into multiple responses and responsibilities, both theo-retical and ethical, in Wolfe's terms—and extinction is one of the gravest challenges to ethics and action confronting all of us today, even if it does so in varied and diverse ways. The reverberations of our inadequacies are everywhere apparent.

The modes of responsibility advocated here also offer, we believe, a more modest, more earthly, and more mature response to the current mass-extinction event than is frequently taken. Faced with species loss, climate change, and the advent of the

Anthropocene, or simply with the march of entropy, actions and stories all too often revert to comfortable, all-too-human scripts, whether heroic epics of conservation and, increasingly, resurrection and de-extinction, enamored of the salvific power of scientific control (Turner 2007), or nihilistic fables of fated human escape or disappearance, erasing all unique forms of nonhuman value in narratives of accelerated progress or inevitable decline. Our response to extinction forgoes these temptations. By *staying with* the lives and deaths of particular, precious beings; by refusing to allow the perspectives afforded by evolutionary deep time or genetic codification—invaluably unsettling as they are—to invalidate the fragile temporalities by which singular living communities make their worlds and make their way in ours; by *holding together* the agencies of different animal species and those of human actors, from biologists to local residents, as well as the institutions, narratives, economies, and motives that give shape and limits to these various, entwined potencies and potentialities; by acceding first *to witness* the multifaceted, and neither inevitable nor (entirely) reversible, unraveling of forms of life that is occurring around us, before yielding to either confident solutions or despairing submission; that is, again, by *staying with* the lives and deaths of particular, precious beings—in these ways, we hope to open up a place and a moment for a reflective gathering of energies *against* extinction but also *creative* of new modes of survival and fragile flourishing, of solidarity and respectful separation, new earthly webs of biocultural prosperity among the wounded and unloved, the precarious and the ruined.[7]

In the course of our collaborative work, three key concepts have appeared and reappeared and have, in various ways, shaped much of our analysis: time, death, and generations. Extinction is fundamentally a deathly process. It is, by definition, a collective death, the end of a living kind. But this larger ending is pieced together out of the deaths of countless individual organisms. In a range of different ways, at various scales, death is central to extinction processes. Many of these deaths happen violently: a Honshu wolf shot and collected as a specimen, a bird of paradise sacrificed for fashion, a Hawaiian monk seal bashed to death on a beach, and countless Passenger Pigeons killed for almost any conceivable reason (chapters 1, 3, and 4, and afterword). In other instances, these deaths are the product of more diffuse processes of change and loss, like the decline of habitat and the impact of fishing and other extractive industries (chapters 3, 4, and 5 and afterword). And then sometimes death is a necessary or, perhaps, not so necessary part of efforts to prevent extinction (chapters 2 and 6).

As a collective death, however, extinction fundamentally demands attention to generations. To understand what is lost in extinction, we must come to terms with species as intergenerational heritages. The significance of extinction, what separates it from the singular death of an organism, is precisely this: the ending of an ongoing lineage cultivated over hundreds of thousands, perhaps even millions, of years of evolutionary time; the abrupt termination of a whole way of life, a mode of being that will never again be born or hatched into our world. As we have explored elsewhere, every species is the unique "achievement" of long lineages of life in which countless generations have each brought forth the next, gifting them, through complex processes of biocultural inheritance, both a material form and a form of life (Rose 2006, 2011, 2012). Extinction is the irreparable disruption and destruction of the *generativity* of such generations.

Finally, thinking extinction has frequently drawn us into consideration of temporalities: from the deep-time processes of evolution and speciation, to the frighteningly rapid pace at which biodiversity loss is today taking place. The ways in which particular species make their lives depend on distinctive and often fragile synchronies and patterns, speeds and slownesses, interwoven temporalities increasingly interrupted by the disturbances of a species "out of time," pursuing short-term profits or producing near-immortal products (chapters 5 and 6). Yet the ways in which we might study and, indeed, try to counter extinction draw too on the distinctively multiple temporalities of storytelling, on creative attempts to produce new ways of understanding and relating to time, of measuring and counting time, of *taking* time—ours and theirs—and of giving it back to creatures prematurely deprived of the time they need to prepare their own resilient generations, to face their own fruitful deaths.

Thus we see that time, death, and generations are, of course, inextricably tied together, with and against extinction. Death, and the relationship of the living to the dead, is a necessary part of the intergenerational production and transmission of ways of life, of the instincts and cultures, the skills and knowledges, by which differently evolved animalities are able to *be*—that is, to create their worlds. These singular, open worlds are bequeathed and haunted by their own unique inheritance, which specifies and stimulates their ongoing potentialities (Lestel, Bussolini, and Chrulew 2014). Reckoning with this gift of time and the responsibility and reciprocity that it engenders is the task of the living, but it is something that can also become so ruptured, so disturbed, that the ongoing relationship to the dead and to time, the task of generation, can no longer be sustained (Derrida 1992; Rose 2006, 2012).

Structure: Questions and Challenges

Each of the chapters assembled here focuses on a particular endangered or extinct *animal* species. While many of the contributors to this book, along with other thinkers on extinction, also conduct research on species in the other kingdoms of life, such as plants or fungi, or on the loss of human languages and other forms of cultural diversity (Sodikoff 2012), we have maintained a central focus on animals and their ways of life in an effort to create a more cohesive collection. It is not our intention to limit extinction studies to a consideration of animal species, and we are confident that those concerned with botanical, fungal, bacterial, and other forms of life, and other ways of living-with various nonhuman (and perhaps nonliving) others, as well as cultural or linguistic survival, will find much in these pages to inform their work on living and dying in multispecies communities.

The book is conceptually divided into two sections. The first three chapters call us into questions. They take inspiration from James Hatley's remarkable observation that extinction is a "disappearance [that] not only is to be questioned but already is a questioning, uncannily interrogating we who remain behind."

Hatley invites us to join him in contemplative walking along a pilgrimage route in Japan that transects the former range of the extinct Honshu wolf: Ōkami. Rejecting easy notions of absolute loss, Hatley investigates the ongoing life of Ōkami, contributing to that life at the same time in his writing. Beyond the meaning of extinction, Hatley enjoins us to consider questions of witness, memory, attentiveness to absence, and the fragility of life not only in our era of loss, but always.

Matthew Chrulew draws us into the interface between the zoo and the wild, introducing us to golden lion tamarins in Brazil and the carers who try to prepare captive animals for life beyond the cage. The story of how scientists collaborated with GLTs to save the species from extinction tells of an experiment unfolding within the lives of the tamarins as scientists learned through trial and error. It is a story of fragments and renewal, filled with ethical dilemmas, miscalculations, sudden and terrible deaths, significant learning, equally significant meta-learning, emotional investments, and boundary-breaking methodologies, that shows how "new natures, cultures, animalities, and subjectivities" are conceived and created "amid overwhelming and singular losses."

Rick De Vos offers encounters with Elliot's Bird of Paradise in Papua New Guinea, bringing us into a shadowy zone where cultural history, colonization, the plume boom, and science converge with indigenous knowledge and hunting. Human–bird relationships of intimacy, mimicry, and admiration are overtaken by commercial hunting, and the question of what is lost is overtaken by science to become a question of taxonomy. As colonization and concepts of hybridity write over extinctions, and write over ways of living, Elliot's Bird of Paradise becomes ever more penumbral, questions about it ever more fraught with mystery.

The last three chapters address action and attention across generations and species. The challenge is described by Thom van Dooren: "the work of holding open the future and responsibly inheriting the past requires new forms of attentiveness to *biocultural* diversities and their many ghosts."

Deborah Bird Rose takes us to the Hawaiian Islands to encounter critically endangered monk seals and the volunteers who protect them on the beaches. She explores the various trajectories toward life and death found within different monk seal populations. Her investigation includes, but moves beyond, the scientific community, with its rational outcomes. She offers an account of communities of otherness within which arise multispecies surprises, prayers, and blessings.

Michelle Bastian asks questions about time. She interrogates proximity, ethics, synchrony, and asynchrony with the aim of challenging us toward considering more complex, shifting, multitudinous understandings of time. Her exploration of time in the life world of endangered leatherback turtles leads her not only to beaches and nests, but to jaguars, jellyfish, fishermen, plastic bags, scientists, and conservationists in both the Pacific and Atlantic Oceans. She offers us a radical call to decentralize clock time and to think and care in multiple temporalities.

Van Dooren invites us to consider inheritance in the context of the critically endangered Hawaiian crow, the forests it once inhabited, the people who seek to prevent its extinction, and the people who oppose measures to reintroduce and protect free-living crows. His focus is on the twinned moment of inheritance: the continuities and the transformations in the context of multispecies biocultural heritages.

Finally, in a lyrical afterword, Vinciane Despret offers a poetic exploration of loss and the question of mourning. Questions of who mourns, and why, are approached within the understanding that "the entire universe thinks and feels itself." Her profound challenge to the implicit question of mourning brings us deeply into the immensities of destruction and grief amid the proliferation of animal worlds.

NOTES

1. For more information on this group, please see Extinction Studies Working Group, www.extinctionstudies.org.

2. For a discussion of the way in which nonhumans might write their own stories in/on the landscape, as well as in humans and our stories, see Benson (2011).

3. For a discussion of the way in which some nonhumans—in this case, Little Penguins and flying foxes—"story" their places/worlds, see van Dooren and Rose (2012).

4. Donna Haraway (2014) has made this point in conversation with Marilyn Strathern's (1992:10) work on the ideas we think other ideas with.

5. Our particular approach takes inspiration from James Hatley's (2000) work on narrative and testimony in the face of the Shoah. Hatley forcefully reminds us of the ethical demands of the act of writing,

of telling stories. In place of an approach that would reduce others to mere names or numbers, in place of an approach that aims for an impartial or objective recitation of the "facts," Hatley argues for a mode of witnessing that is from the outset already seized, already claimed, by an obligation to those whose story we are attempting to tell. In the context of ecocide and many other forms of mass death, this mode of storytelling is particularly important. Along with the imperative of remaining true to the facts of the situation, witnessing insists on truths that are not reducible to populations and data, a fleshier, more lively truth that in its telling may draw others into a sense of accountability and care.

6. For further insight on this broad collection of approaches, see van Dooren et al. (2016).

7. As Mick Smith (2013) puts it: new forms of posthuman ecological community amid the loss of irreplaceable ways of being in the world. This approach to "staying with" draws on, and is in conversation with, Haraway's (2016) work on "staying with the trouble."

REFERENCES

Apffel-Marglin, Frédérique, and Stephen F. Marglin, eds. 1996. *Decolonizing Knowledge: From Development to Dialogue*. Oxford: Clarendon Press.

Barnosky, Anthony D., Nicholas Matzke, Susumu Tomiya, Guinevere O. U. Wogan, Brian Swartz, Tiago B. Quental, Charles Marshall, Jenny L. McGuire, Emily L. Lindsey, Kaitlin C. Maguire, Ben Mersey, and Elizabeth A. Ferrer. 2011. "Has the Earth's Sixth Mass Extinction Already Arrived?" *Nature* 471:51–57.

Benson, Etienne. 2011. "Animal Writes: Historiography, Disciplinarity, and the Animal Trace." In *Making Animal Meaning*, edited by Linda Kalof and Georgina M. Montgomery, 3–16. East Lansing: Michigan State University Press.

Buchanan, Brett, Jeffrey Bussolini, and Matthew Chrulew. 2014. "General Introduction: Philosophical Ethology." *Angelaki* 19, no. 3:1–3.

Chrulew, Matthew, and Rick De Vos. Forthcoming. "Extinction." In *The Edinburgh Companion to Animal Studies*, edited by Lynn Turner, Undine Sellbach, and Ron Broglio. Edinburgh: Edinburgh University Press.

Colebrook, Claire. 2014. *Death of the PostHuman: Essays on Extinction*. Vol. 1. London: Open Humanities Press.

Derrida, Jacques. 1992. *Given Time*. Vol. 1, *Counterfeit Money*. Translated by Peggy Kamuf. Chicago: University of Chicago Press.

De Vos, Rick. 2007. "Extinction Stories: Performing Absence(s)." In *Knowing Animals*, edited by Laurence Simmons and Philip Armstrong, 183–195. Leiden: Brill.

Griffiths, Tom. 2007. "The Humanities and an Environmentally Sustainable Australia." *Australian Humanities Review* 43.

Haraway, Donna. 1991. "Situated Knowledges: The Science Question in Feminism and the Privilege of Partial Perspective." In *Simians, Cyborgs, and Women: The Reinvention of Nature*, 183–202. New York: Routledge.

——. 2014. "SF: String Figures, Multispecies Muddles, Staying with the Trouble." Lecture presented at the University of Alberta, Edmonton, March 24.

——. 2016. *Staying with the Trouble: Making Kin in the Chthulucene*. Durham, N.C.: Duke University Press.

Hatley, James. 2000. *Suffering Witness: The Quandary of Responsibility After the Irreparable*. Albany: State University of New York Press.

Human Animal Research Network (HARN) Editorial Collective, ed. 2015. *Animals in the Anthropocene: Critical Perspectives on Non-Human Futures*. Sydney: Sydney University Press.

Ingold, Tim. 2013. "Anthropology Beyond Humanity." *Suomen Anthropologi* 38, no. 3:5–23.

Kingsford, R. T., J. E. M. Watson, C. J. Lundquist, O. Venter, L. Hughes, E. L. Johnston, J. Atherton, M. Gawel, D. A. Keith, and B. G. Mackey. 2009. "Major Conservation Policy Issues for Biodiversity in Oceania." *Conservation Biology* 23, no. 4:834–840.

Kirksey, S. Eben, and Stefan Helmreich. 2010. "The Emergence of Multispecies Ethnography." *Cultural Anthropology* 25, no. 4:545–576.

Kohn, Eduardo. 2013. *How Forests Think: Toward an Anthropology Beyond the Human*. Berkeley: University of California Press.

Kolbert, Elizabeth. 2014. *The Sixth Extinction: An Unnatural History*. New York: Holt.

Lestel, Dominique, Florence Brunois, and Florence Gaunet. 2006. "Etho-Ethnology and Ethno-Ethology." *Social Science Information* 45:155–177.

Lestel, Dominique, Jeffrey Bussolini, and Matthew Chrulew. 2014. "The Phenomenology of Animal Life." *Environmental Humanities* 5:125–148.

Lorimer, Jamie, and Clemens Driessen. 2014. "Wild Experiments at the Oostvaardersplassen: Rethinking Environmentalism in the Anthropocene." *Transactions of the Institute of British Geographers* 39, no. 2:169–181. doi:10.1111/tran.12030.

Myers, Norman, and Andrew H. Knoll. 2001. "The Biotic Crisis and the Future of Evolution." *Proceedings of the National Academy of Sciences* 98, no. 10:5389–5392.

Raup, David M., and J. John Sepkoski. 1982. "Mass Extinctions in the Marine Fossil Record." *Science* 215, no. 4539:1501–1503.

Rose, Deborah Bird. 2006. "What If the Angel of History Were a Dog?" *Cultural Studies Review* 12, no. 1:67–78.

——. 2011. *Wild Dog Dreaming: Love and Extinction*. Charlottesville: University of Virginia Press.

——. 2012. "Multispecies Knots of Ethical Time." *Environmental Philosophy* 9, no. 1:127–140.

Smith, Mick. 2005. "Hermeneutics and the Culture of Birds: The Environmental Allegory of Easter Island." *Ethics, Place and Environment* 8, no. 1:21–38.

——. 2013. "Ecological Community, the Sense of the World, and Senseless Extinction." *Environmental Humanities* 2:21–41.

Sodikoff, Genese Marie, ed. 2012. *The Anthropology of Extinction: Essays on Culture and Species Death*. Bloomington: Indiana University Press.

Star, Susan Leigh. 1990. "Power, Technology and the Phenomenology of Conventions: On Being Allergic to Onions." In "A Sociology of Monsters: Essays on Power, Technology and Domination," edited by John Law. Special issue, *Sociological Review* 38, no. S1:26–56.

Strathern, Marilyn. 1992. *Reproducing the Future: Essays on Anthropology, Kinship and the New Reproductive Technologies.* New York: Routledge.

Tsing, Anna. 2011. "Arts of Inclusion, or, How to Love a Mushroom." *Australian Humanities Review* 50:5–22.

Turner, Stephanie S. 2007. "Open-Ended Stories: Extinction Narratives in Genome Time." *Literature and Medicine* 26, no. 1:55–82.

van Dooren, Thom. 2014. *Flight Ways: Life and Loss at the Edge of Extinction.* New York: Columbia University Press.

van Dooren, Thom, Ursula Münster, Eben Kirksey, Deborah Bird Rose, Matthew Chrulew, and Anna Tsing, eds. 2016. "Multispecies Studies." Special issue, *Environmental Humanities* 8, no. 1.

van Dooren, Thom, and Deborah Bird Rose. 2012. "Storied-Places in a Multispecies City." *Humanimalia: A Journal of Human/Animal Interface Studies* 3, no. 2:1–27.

——. 2016. "Lively Ethography: Storying Animist Worlds." In "Multispecies Studies," edited by Thom van Dooren, Ursula Münster, Eben Kirksey, Deborah Bird Rose, Matthew Chrulew, and Anna Tsing. Special issue, *Environmental Humanities* 8, no. 1:77–94.

Whatmore, Sarah. 2003. "Introduction: More Than Human Geographies." In *Handbook of Cultural Geography*, edited by Kay Anderson, Domosh Mona, Steve Pile and Nigel Thrift, 165–167. Thousand Oaks, Calif.: Sage.

Wolfe, Cary. 2003. *Animal Rites: American Culture, the Discourse of Species, and Posthumanist Theory.* Chicago: University of Chicago Press.

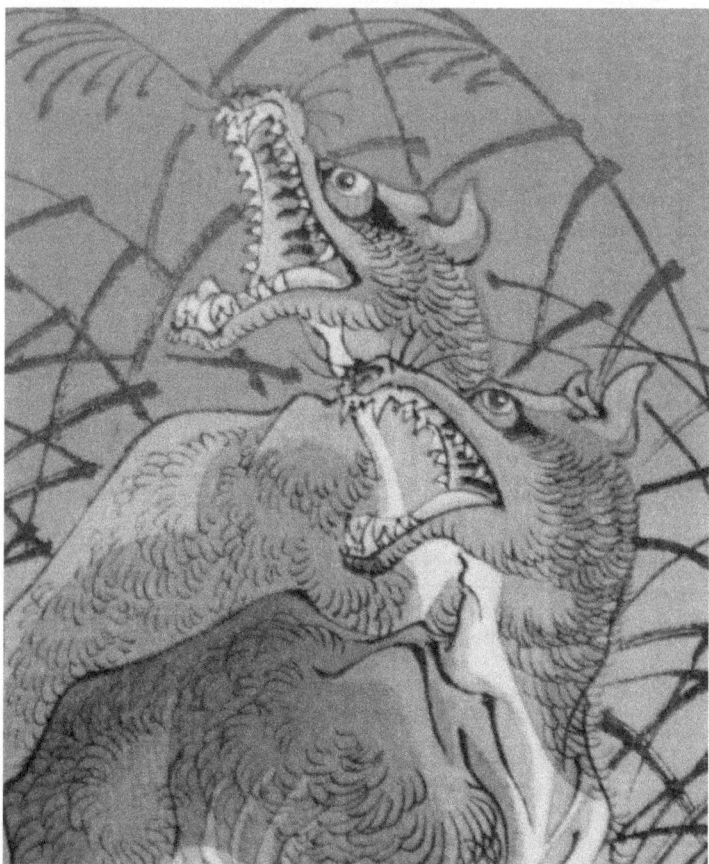

Tsukioka Yoshitoshi (1839–1892), *Kitayama Moon: Toyohara Sumiaki (Kita-yama no tsuki: Toyohara Sumiaki)*. (Detail from number 32 of *One Hundred Aspects of the Moon* [*Tsuki hyakushi*, 1886]. Woodblock print, oban tate-e, 13⅞ by 9½ in. [35.2 by 24.1 cm])

1. WALKING WITH ŌKAMI,
THE LARGE-MOUTHED PURE GOD

JAMES HATLEY

> I walk
> with that wolf
> that is no more.
>
> Toshio Mihashi

JANUARY 1

Making my way in enforced somnolence over the Pacific, I have finally worked up the courage to crack open the flap on my peep-hole-size window, the one that I and my fellow passengers had been specifically instructed by flight attendants to keep firmly shut. Below, in the half-light of winter, an ocean disappears into thickening shadows of blue, as a thin layer of pink clouds stretches along the vast curvature of Earth's horizon. The sky is taking in its last breath of light before nightfall. Earth might be, against the time scale of billions of years, a doomed project, a fleeting apparition, but at the moment it is something quite other than flimsy. The opening verses of Psalm 19 come to mind:

> The heavens declare the glory of God;
>   and the firmament showeth his handiwork.
> Day unto day uttereth speech,
>   and night unto night showeth knowledge.

Even if I might not discern knowingly what is being proclaimed as the boundary between darkness and light advances across the face of Earth, I would be a fool to believe that speech and understanding are not already at work out there from the stratosphere down to Earth's core, speech and understanding most often beyond my ken. Buddha mind, Kūkai might add, is busy preaching.[1]

After a three-hour delay in San Francisco, many fitful naps, and a steady diet of canned movies and scratchy audio, my four students and I soon will be descending toward the Kii Peninsula on Honshu, the largest of the islands that make up Japan, where the evening of January 2 is already setting in. Even as the stony paths of the Kumano Kodō await our first steps, the pilgrimage has already begun, my third in three years. This time, my trek is to be dedicated to the memory of the Honshu wolf, last seen—in fact, last killed—on January 24 or 23, 1905, not so far from the trail we will be walking in the Kii Mountains. His Japanese name was Ōkami, sometimes rendered in a creative etymology in Japanese as "the Great God" (Walker 2005:9). Scientists have come to know him as *Canis lupus hodophilax*. For the rest of the existence of this planet, on each and every day, Ōkami will have been extinct.

JANUARY 3

Orientation Day. Dr. Kumi Kato, our trip leader, reminds us that we are walking the Kumano Kodō in a time of natural catastrophes that have been, in turn, intensified by human negligence and environmental malpractice. The earthquake and tsunami of the previous March on the northern end of Honshu registered here in the south as an imperceptible tremor on the Kii Peninsula, followed a few hours later by several gentle swells lapping against the shoreline of Wakayama Bay. But those waters nevertheless communicated disaster, one that has thrown the entire country into deep reflection about its relationship to the natural world, a relationship that has had its destructive moments on both sides throughout Japan's history. As if to underline this fact, a powerful cyclone hit the Kii Peninsula in September, precipitating landsides and massive flooding across the mountainous landscape. The Shinto temple at Nachi Falls, the final destination for our own pilgrimage along the Kumano Kodō, was buried 9 feet deep in boulders and mud. Several hundred died in the mayhem, Kumi tells us. In one deep mountain valley on the Tonda River, the crest of water rushing downstream wailed through the storm like a crying baby. "As if the land itself were speaking," an inhabitant of the mountains would relate to us later. "The tsunami taught me the meaning of a community of hands," Kumi shares with us. "In our modern culture we are used to machines taking care of our lives but after the tsunami the machines were drowned or stranded. Only hands were left. With just our hands we cleaned up a village."

JANUARY 4

Today finds us at Takijiri-oji, entryway into the most revered stretches of the Nakehechi route of the Kumano Kodō. The walking begins. After rinsing our hands in a stone pool of flowing water, we approach a shrine, leave an offering, clap our hands to alert the gods inhabiting this place, pull on a hanging rope that rings a bell, and bow. Beyond, a narrow trail climbs upward across great stones. Scaling the steep ridge of Tsurugi-no-yama (Sword Mountain), we labor in the cold morning air. That the opening ascent is so daunting is no accident. Jiso, both grocer and docent for the heritage site, has exhorted us in words translated by Kumi and reminiscent of the Buddha's first noble truth: "Your suffering, bring it on the trail. Let it afflict you, so that it can be washed away afterward." Even as this thought comes to mind, the deep, throbbing notes of a primitive trumpet boom into the mountains from the gorge below and stop me in my tracks. I realize that Jiso is sending the pilgrims off on their journey with a final exhortation. His gift of breath, focused and magnified through the capacious hollow of a section of bamboo, is chimeric, something both human and more than human. The stones of the mountain slope absorb the sound's energy and transform it, in that manner peculiar to stones, into intensified silence and attentiveness. "They are listening, too," I whisper to myself. Earlier, Tempei Miyaji, a Japanese colleague and Kumi's former student, had explained how slats of wood hanging on the wall in the Kumano Kodō Kan Pilgrimage Center and inscribed with calligraphy were each a prize-winning local haiku. I asked if he might translate one. "This is difficult. So many senses are being played out in the Japanese," he reported, "so impossible to explain." Still he politely offered these words sketching out a first rendering of one of the poems:

Each stone has its purpose—
boats of prayer
setting out

Silence, even a silence that is of stone, is not inert. It, too, speaks. Stone speaking to stone, as day is speaking to day, and night to night. The diminished sunlight, the icy patches of recently fallen snow, the mingling of wind and trees—all these things and all the other 10,000 things, at this moment, in this place, instantiate Buddha mind and offer a sutra of compassion, a hand of healing, and a persistent invitation to enlightenment. Might not, then, the entirety of existence in all its manifold entities and elements indeed be a boat of prayer composed of boats of prayer? I am reminded of a passage from Kūkai (2004), the founder of the Japanese sect of Shingon Buddhism, whose traditions permeate the pilgrimage path we have now begun to travel: "Any sound [uttered] when one first opens the mouth is accompanied by the sound *Ah*, and without the sound *Ah* there would be no speech at all. Therefore it is deemed to be the mother of all sounds" (107). Has not Jiso's final word to us, shorn of all consonants and utterly consumed in the sound of breath reverberating in its breath, been precisely this sound?

The world speaks in a plethora of tongues. Might Ōkami's be among them still?

> On the barren tree
> enough leaves have remained to
> populate the wind.[2]

> Trunk upon tree trunk
> cracks! Wayfarers called to prayer through
> veils of cascading snow.

Even after several years of reading about the religious practices and history permeating this landscape, I know so little about what actually surrounds me here. Not only because I find myself in a strange country, an outsider to its traditions and language, but also because, like most citizens of the allegedly developed world, I am embarrassingly ignorant of the specific flora and fauna of the living Earth *wherever* I might go. Thankfully, at least a few species are familiar enough that I can name them confidently. These are the ones, I note ruefully, that have been assiduously transplanted into the gardens of North America and then have escaped to grow willy-nilly across its landscape. But they appear here and now in their very wildness, at ease on a mountain slope, rooting in effortlessly among dark stones—Suzuki grass and cedar, cypress and bamboo. In particular, I cannot help but become entranced by the abundance of camellias blooming self-assuredly along the frigid pathway as it levels off and then more gradually climbs toward Takahara, the site of two great camphor trees and our resting place for the first night on the trail.

> Even as I pressed
> fallen petals between my fingers,
> their odor fled.

January 5

> With each step, the howling of wolves is not there . . .
> With each step, the howling of wolves is not there . . .
> With each step, the howling of wolves is not there . . .

Something has changed. Today began with clouds blowing in over the mountains, as falling snow turned the landscape white all the way down to the floors of the deepest valleys. And now I have stopped along the Kumano Kodō at a clearing where a deserted teahouse is succumbing to the forest. Ghosts are here, or at least the intimations of ghosts.

> Near a door unhinged—
> snowflakes settle into the hollow
> of a charred, fallen beam.

The silence permeating this site redoubles the silence of the forest surrounding it. The inn's hosts have departed, their greetings no longer offered to wayfarers laboring up this mountain gorge. No more can human voices be heard murmuring, lost in leisurely conversation amid the songs of birds, the stirring of leaves on the forest floor, and the steaming of a tea kettle in the kitchen. Now only termites and ants, voles and field mice, crows and quail are making themselves at home. Increasingly, all that remains of this phantom dwelling is the thought of what it once was. But who remains then to think it?

Looking again into the forest interior, I am struck by how the sun's wintery, tentative illumination, weakened further by the gray clouds overhead, is nevertheless effortlessly gathered into and kept alive by a glistening skin of snow coating every branch and leaf in sight. The dying embers of late-afternoon light are conserved and distilled, as a world of crystalline luminosity temporarily overturns the inevitable descent into a night of rough bark and rotting stump, fallen leaf and moldy earth.

Yet this ethereal light does not warm the blood, even if it entices the eyes. And the blood cannot be ignored. It would be unwise to expect the flesh to dissipate without a fight. Shivering is called for. And words as well that are true to the blood.

Seeing wolf shit
these weeds feel
even colder.

Kobayashi Issa (Greve 2006)[3]

In another time, the poet might have written these lines in this very place. But with each step I take, neither howling nor scat nor scent of wolf is anywhere in evidence. And so another kind of ghost draws near: that of a living kind who never again will roam these forests or any other like them.

Perhaps the time has finally come in this journal to speak more straightforwardly of the disastrous loss of Ōkami, which is to say, in a word, of her extinction. "Extinction . . ." As I sound out its syllables, my breath condenses in the frosty air before me. Word becomes vapor, and then vapor dissipates into silence. Only the surrounding trees and stones offer confirmation, such as they are able, that anything might have been uttered. And I wonder: What's in this word and its being spoken here that has to do with Ōkami? "Extinction" is that term in which we humans, who would opine endlessly about the living kinds, announce a grave new concern for them, a concern we have only recently, in the past three centuries of human existence, even been able to conceptualize. And yet the very idiom by which this concern is voiced all too often proves itself to be antiseptic and distant. "Extinction" is the word one uses when one discusses policies and lists, when one determines dates and definitions. "Mass extinction," for example: How many species must no longer be dancing on the surface of this planet, if not on the head of a pin, before singular and straightforward extinction accumulates into its mass variety? It's a good question, and the thinkers of extinction do well to worry about it. But is it the right question for this moment on this trail? Yet another salient point: If the fossil record is consulted, we find that close to 100 percent of all species that ever existed are now extinct. Surprisingly, the state of extinction proves to be the very norm of species existence, as each living kind emerges, flourishes, and then withers away on the vine for one reason or another. *Canis lupus hodophilax* has now joined this supermajority. One might also worry about the ontological status of an extinction that is of a subspecies rather than an entire species. *Hodophilax* has departed but *lupus* remains.

But all this seems peripheral to the question at hand, whatever that question might turn out to be. For *there is* (*es gibt*) the loss of Ōkami as a species, and with this realization comes a sense that some questions must be posed. Indeed, one is called on to witness that Ōkami's disappearance not only is to be questioned but already is a questioning, uncannily interrogating we who remain behind. Is "extinction" even the right word to turn to in order to respond to

Ōkami's disconcerting invitation to thought? Yet if not this word, then what word or words are to be offered in the circumstances of Ōkami's disastrous excision from the living? Indeed, what is/are the specific questions I am being called on to ask, as this trail is followed and this journal is written and then rewritten even years later. Somehow this loss, irrevocable and terrible, is to be registered in some fashion, however wayward, for some reason, however ephemeral.

Something like Ōkami should not disappear in our time without its departure having been noticed, without its loss mattering. This particular thought registers viscerally. And it calls for a mode of philosophical reasoning that takes one's gut seriously. And takes Ōkami's interest in the matter seriously as well. A question of etiquette is involved, "transhuman etiquette," as Anthony Weston (1994:145–167) would put it.

The discourse of species extinction demands its specimens. Locked away in some recess of the Natural History Museum in London is a Honshu wolf zoo-ombie, the preserved cadaver collected on January 23, 1905, by Malcolm Anderson, an American agent for the Duke of Bedford, an English zoologist. The wolf was shot not so far from here on the Kii Peninsula in the neighborhood of a small village now named Higashi Yoshino. The specimen's demise serves as the last moment known to science of *Canis lupus hodophilax*, or Ōkami, having lived in these woods. The specimen wolf has evidently turned out to be the last one of its kind known to humankind. And so, for now, it's more or less official: Ōkami is extinct.

The very thought of that specimen haunts me. And shames me. I have seen pictures of his preserved kin on the Internet, mangy and stiff, empty eye sockets filled by lifeless glass orbs. A stuffed wolf is posed by the taxidermist as if ready to leap forward but looks in truth more likely to scatter into pieces if it's even touched. The teeth seem ready to fall out of their desiccated gums. The specimen surely has become prey to dry rot and home to at least a few insects looking for a handy place to mate and nest. And yet we humans want to look at it, look at it up close, in its stuffed skin. Our eyes are hungry for confirmation. And if a specimen is not nearby, then we are content to stare at its image in all its shabbiness on the Internet. Colonial exploration, Mark Barrow (2009:5–6) reminds us, fueled a grand era of specimen collecting that, in turn, laid an important part of the groundwork for the emergence of species biology and the theory of Darwinian evolution. Mirroring Freud's dueling notions of thanatos and eros, the drive of the colonial mind to search out and accumulate examples of diverse living kinds across the face of Earth found both creative and destructive instantiations (Barrow 2009:63–64, 153–166). Often these opposing elements were inextricably mixed together, as natural historians hurried to collect as many living individuals as possible of a rare species, thus inevitably becoming a contributing cause to its very extinction (De Vos, chapter 3, this volume). But was the rush to extinction so inadvertent? The panopticon of earthly kinds being constructed by an emerging scientific enterprise of global biology demanded exemplary objects on which to gaze. That we humans had verified a species to have existed was deemed more important than the fact that it might continue to exist without our having known it as doing so. As a result, the face of Earth became in principle the planetary backroom for a great and celebrated complex of museums, botanical gardens, and zoos. This is, it strikes me, not a very polite manner in which to conduct oneself in the presence of all the other living kinds.

The Kumano Kodō as an aisleway in a storeroom of living kinds! This, I think whimsically, darkly, constitutes the very form of the Anthropocene, of an Earthscape diminished both in its living kinds and in its living significance by a pervasive regime of anthropogenic extinction. Extinction is a necessary thought, but it is not the best thought in regard to Ōkami here and now.

Walking the trail, I find myself immersed in solitude. As one step is taken, another arises. And so another, another, and another. Ridge comes into view, and ridge is left behind. Imperceptibly, the mind lets go of itself and begins to wander in places its alleged owner does not recognize. My gaze is increasingly absorbed in the patterns of shadow and light playing among innumerable ferns, playing across the rough, frizzy bark of cedar trunks, playing in the glint of snow on the trail.

Until, suddenly, pacing near me is the king of the Hokkaido wolves. Not the smaller Honshu wolf, only 1 foot tall, which inhabited these woods, but the Ezo wolf (*Canis lupus hattai*), the extinct wolf of Japan's northernmost island. "Really?" I wonder. "Where did this thought come from?" Then, "A king of the 'kindom' of the departed, a revered ghost," I think. "Here I am," I offer. Now the wolf is by my side. Panting. Suggesting that I follow him into the mountains. He is large and muddy. He has come a great distance to visit me. "Do you know how it was to lope along these hills, to be ruled by their scents, to lock one's teeth on the throat of a deer?" he asks me. For a while, he paces with me. And then he is gone. "Why was I thinking these thoughts; why was this presence so palpable?" I wonder. "I didn't exactly see anything, and yet he was there, wasn't he?" Was this a moment of self-initiated psychosis or a visitation? Or both? Or neither?

The abyss in time left by Ōkami's extinction works on me uncannily. The trail's invitation to meditation is not without efficacy, however unsettling it might be.

Winter mountains,
and no passage given:
snout-to-snout with a wolf.

Masaoka Shiki (Greve 2006)[4]

At this moment a question must be asked of the thinker who thinks he has been named in a visitation by a ghost of one species to walk with the ghost of another. The question has to do with how these thoughts of a living kind, extinct or no, have taken a life of their own that is purportedly not merely the working of an argument or a play of the imagination. The animals are pacing, the thinker swears, here and now in the words of this sentence. Palpably so. How is the contempoary modern or postmodern critic to make any sense of that? Is it not unseemly to ask one's philosophical interlocutors to attend to what comes from a questionable visitation in such questionable terms? A question of etiquette, of etiquette both modern and postmodern, is involved here. Thinkers don't speak of their dreams in public. Many, I suspect, do not even speak of their dreams to themselves.

No answer emerges to these doubts that would be straightforwardly satisfying. But what does come to mind is that in an era of anthropogenic species extinction arguably fueled by an ideology of hyperbolic doubt, perhaps hyperbolic affirmation, or what could also be termed uncanny affirmation, is also being called for. In a time of mass species extinction, to speak healing words in proximity to the living Earth becomes paramount. In Arrernte country of Australia's "Red Centre," Margaret Kemarre Turner (2010) intimates, the affirmation of her land demands a story affirming its beauty: "If you're seeing the Land without the Story, then there's nothing there. We see our country, even though it might be destroyed by another species, we see how the beautiness is still in the country. . . . It doesn't matter that horses and bullocks have caused such destruction, we still see the spirit of the Land glistening" (126–127). Kemarre Turner's suggestion opens up a way to philosophical discourse that begins not in negation and critique but in affirmation and amplification, an approach common to the wisdom traditions of many indigenous cultures. In a gesture similar to that of Kemarre Turner, philosophers Freya Mathews and David Abram have appealed to Spinoza's philosophical legacy in order to argue for pan-psychism (Mathews 2003:48) or pan-intelligibility (Abram 2010:109) as a fruitful approach to reorienting philosophical reasoning to the manifold yet specific modes of significance at work in our habitation of singular places. Abram (2010:119–20) insists that the philosopher think anew the ancient Greek insight that species and idea are deeply interconnected, an interconnection that is rendered palpable by attending carefully and patiently to a unique place under the sun. This, in turn, leads to modes of philosophical discourse nested from the beginning in discursive practices of appeal and etiquette rather than in those of argument and dialectic.

In death, the remains of a wolf—fangs, pelts, and skulls—were preserved and hung up as talismans to ward off evil spirits and, in some cases, treated as objects of worship (Knight 2006:206). Even the Honshu wolf's scientific name, *Canis lupus hodophilax*, assigned by Coenraad Temminck in 1839, is indicative of this wolf's vocation as protector. *Hodo* means "path" or "way" in Greek, whereas *philax* means "guardian." So the Greek suggests something like "guardian of the way." Folk tales and personal anecdotes abound of *okuri Ōkami*, an "escort wolf" accompanying home lone walkers on a dark forest path (Knight 2006:205). Perhaps this was a predator contemplating the status of its possible prey? Or perhaps the wolf was just interested in sharing a walk. Beyond those easy answers lie yet other possibilities. We of the Anthropocene, who are in the habit of knowing the living world from afar, underestimate the crazed moments of sociality and reciprocity emerging between creatures, both human and more than human, in intimate, even if contested, daily contact with one another.

In the extinction of Ōkami, humans are left with the remnants of a species' one-time existence on the face of Earth—some preserved specimens, some talismans, a few poems, and a scattering of images. Also, shrines and stories, rituals and prayers. And at the center of all these: the name of Ōkami.

Ōkami. This name, with all its attendant names, both haunts and guides me. Indeed, it is through the name of Ōkami that I first even came to know of its storied existence on Earth. Embarrassingly, even shamefully, I started this journey into extinction by scanning on the Internet a list of vanished species once endemic to Japan. Noticing the names of two wolves and determining by means of Wikipedia that one of them had inhabited the mountains surrounding the Kumano Kodō, I decided that the Honshu wolf, or Ōkami, would become the subject of this journal. Had my finger landed in a different place on that screen, *Hirasea planulata* (an air-breathing land snail), *Lamelli deamonodonta* (yet another land snail), or *Lutralutra whitely* (the charismatic Japanese river otter) might just as well have emerged as my guide. Extinction has become so endemic to our time that choosing (as if choice were the modality by which these responsibilities are to be fulfilled!) which lost species should be remembered, which one or ones should find a place in one's thoughts, has been rendered increasingly arbitrary. And burdensome. Is there enough room in any memory, even that of a museum, to hold the loss of living kinds being undergone at this moment on the face of Earth?

Certainly, better ways must exist to find one's way to an animal, even an animal that has been declared extinct. The very notion that one can put one's finger upon a name on a list and decide, just because, that this is the animal one will think on seems itself indicative of the plight of animals in the Anthropocene—to be surrounded by human beings for whom the perplexity and complexity of the living world has been reduced to an amorphous set of words and a collection of fleeting images. The very practices by which the living world finds its place in human thought is increasingly dominated by a false familiarity in which everything is brought near to the human thinker precisely by its having been first stripped of its manifold living senses and so reduced to a bare minimum of meaning. We are overwhelmed by facile notions of everything and starved for contact with anything real and subtle. I am drawn to this horrifying conclusion: in the very manner the issue of Ōkami's extinction in this journal is being posed, I find myself complicit in the discursive, if not technological, practices characterizing an era of anthropogenic mass species extinction.

Yet these words arising here and now on behalf of Ōkami need not be, I hope, simply the arbitrary repetition of a name on a list. The issue, after all, is not the indication of extinction, as if one were called on to point out authoritatively the extinct other and note its rather regrettable state of nonexistence from afar. Rather, one is called on by the very word "Ōkami," even if one first encountered it as a cipher on a list, to learn to hear its naming responsively and responsibly. The word emerges before any list it might be placed on, before any specimen might present itself as an indication of its biological kind. The naming of Ōkami is in this sense indigenous, belonging in the first instance to those carrying that name. And so, etiquette is involved. "It is not only the vocabulary of science we desire," Linda Hogan (1995) writes. "We are looking for a tongue that speaks with reverence for life, searching for an ecology of mind" (60).

Ōkami's absence in these forested mountains, now a century old, has left me wondering this: In which tone, in what conversation, in whose presence, might now the name of Ōkami be spoken? How might one become worthy of speaking a name that, even in the wake of a living kind's extinction, has not ceased to carry on the report of that living kind? These questions border more on the ethical and the religious than on the ontological and the epistemological. The species of truth emerging here is not one that asks in the first instance that I grasp Ōkami, as if outlining Ōkami's shape for the sake of human knowing were the principal responsibility that Ōkami's extinction now entails. Instead, my very rendering of Ōkami as "extinct" finds that it is already undermined by the ongoing naming of Ōkami, a naming that, in turn, calls me into another sort of relationship with a living kind than I could ever have anticipated on my own. The uncanniness of Ōkami, in both its presence and its absence, in both its emergence and its extinction as a living kind, interrupts my every effort to own the discourse in which I am now speaking: "*Hineni*. Here I am, Ōkami. You have named me."

Innumerable leaves dissipating on the forest floor. For each of these leaves, innumerable prayers are to be offered in innumerable lifetimes. Might, then, each leaf find its own name in a pure land where its dissipation might be finally and uniquely signified? Such dreams dance at the edge of one's consciousness on this pilgrimage path.

JANUARY 6

Have I bitten off more than I can chew? Or worse, is what I have bitten off now chewing on me? I forwent the hike from Hosihmini to Hongu today. Most of the shops are closed for the New Year, so I've wandered around Hongu town looking for a place to sit. I finally found one at a welcoming center with a nicely appointed ice-cream counter and gift store.

Today Andrea, one of the students accompanying me on the Kumano Kodō, woke up with a sore knee, so she stayed in Hongu as well. She is touched by genius, majoring in chemistry, brilliant in math, accomplished in the cello, and in love with Japanese animé. The last I saw her, she was blissfully folding complicated origami on a bench in the center of the temple grounds and then handing them out to passersby. The temple building nearby was damaged during the cyclone and, like the Kumano Hongu Heritage Center, is closed. The magic of the site, one might argue, has been diminished. But all diminishment, even as all intensification, is illusory in the eyes of those who meditate here. And not all, at least for the time being, has been undone by the cyclone. I discovered a small pond that turned out, once I considered it more carefully, to have been constructed as a reflecting pool for an extremely tall camphor growing at one end of it. This watery mirror had been precisely shaped to fit the tree's reflection. Where the trunk stood next to the water, the pool was narrow, but at the other side, where the broader reflection of the tree's crown fell, the pool spread out to take in the entire image.

The effect is uncanny. Two trees—one a ghost, its crown descending into watery depths, and the other a specimen of sap and cellulose, rising toward the winter sun. These two trees are interwoven, as a set of roots from above descends into the other set of roots ascending from below. I return repeatedly to gaze into the pond and up into the sky, and back into the pond once again, until the darkening sky erases the scene.

Carp drifting in water,
floating among golden trees
as evening falls.

The reflection of a camphor tree in the pond becomes more compelling for the moment than the actual tree soaring above it. What Plato consigned to the lowest realms of being in *The Republic* and so dangerously perverting in its influence on thinking, becomes in this pond a most knowing sensei. The moon in water—not only a reflection, but the reflection of a reflection. So many modes and nuances of reality are at work in Buddhism, particularly in tantric traditions such as that of the Shingon lineage, long associated with the Kumano Kodō. Heraclitus, another Greek philosopher, argued that things never stay still or the same. Even when they do. To which a Buddhist might answer that they are, and are not, staying the same. Flow rebounding on flow. There is no manner of fixing or unfixing this whatsoever. And yet here I am speaking of it and in it and by it and through it. Or is it speaking of me and in me and by me and through me? In the spirit of Zhuangzi, the tree dreams the dreamer. And redoubling the dream, the tree in the pond dreams the tree in the air dreaming the dreamer. And the wolf too, extinct, without substance and only shadowy in form, is dreaming all of this and more. Affirmation persists, uncannily and with deference to others, as well as to oneself.

Have we not been too often overconfident of the autonomy of meaning in the Western tradition? Recently in a book by Edward Mooney (2009:53–54), I was surprised and delighted to read his argument, borrowed in part from Ortega y Gasset, that the realities undergirding philosophical tradition are perennially in danger of being lost, of dissipation, in a word, of extinction. The animating reality of the beautiful, the true and the good is not simply there to be had, as if it were stored inertly and in perpetuity in clearly marked cans on a grocery-store shelf of the mind. Instead, the truth of things must be actively and carefully saved from its moldering away by each succeeding generation. The life of the mind, after all, is *alive*. And so can die. And the truth in things can die just as well. Or, even worse, can become moribund and even morbid. Thought is not simply there for the thinking, given enough persistence in rationation on one's own part. Rather, each word emerging here on this page is responding to the silence left in the wake of the words, written and spoken, of Kūkai and Plato, of Ortega and Heraclitus, of Bashō and Bugbee, of Moses and Buddha, and all the rest.

And to the silence left in the wake of Ōkami, too. Even in life, Ōkami was a sensei, a teacher, a philosopher. The many folk tales that sprang up everywhere in Ōkami's shadow and persist beyond

Ōkami's extinction give proof to this. Even now, Ōkami's teaching has not dissipated, so long as a word of Ōkami remains and one is open to hearing it here and now. Is this not finally this journal's theme? Not simply to mourn the loss of Ōkami but also to resist with one's heart, with one's mind, and with one's soul the very loss of this loss, the descent into meaninglessness that extinction of a living kind, particularly anthropogenic extinction, threatens?

Has the very loss of Ōkami's loss as of yet arrived? Not, thankfully, even if also sorrowfully and uncannily, as of yet. This name has not been lost. And to this name, the gut responds and the mind follows.

But the welling up of irretrievable loss and unanticipated emergence that is the very logic of our evolutionary history also instructs us. Extinction, too, is a sensei. Certainly, Heraclitus would have much to say about the subversive dynamics of the human chromosomal structure as it metamorphoses its way through each suceeding generation. And, for this reason, the extinction of the human species is inevitable, is it not? Simply the sheer extent of geologic and zoogenic time in relation to the only very recent appearance of our genotype on the face of Earth by necessity demands that one wonder about our everyday assumptions concerning what matters and what persists. One's place under the sun is temporary. And the place of one's living kind is also temporary. In the face of this realization, one wonders. And the wonder exemplified here is not, I suspect, that of Aristotle, for whom each species, after all, was eternally fixed in its particular form and circumstances. After Darwin, we know better than that.

Imagine our species having existed for 1 billion years. Can we reasonably expect that our children 40 million times removed would still be of our species? Certainly, the biological record would suggest an outcome radically divergent from just more humans until the end of time. Or of more wolves, for that matter. The difference between Ōkami and we humans, I suddenly think, might not be nearly as anarchic, nearly as uncanny, as that between we humans here and now and our descendants 1 billion New Years from now. Unless, of course, the whole line of primates simply dies out, another limb of the tree of life lopped off, perhaps before its time.

The Buddhist perspective on the instability of individual existence finds a cosmological correlate in the doctrine of an infinite number of universes. Not just individuals but whole worlds are ephemeral, relative, empty. Might there also be an infinite number of species, each ephemeral in its own right (which can be said only with the irony inherent in Buddhist thought) in those infinite worlds? The human desire to locate oneself, to know exactly where and on what one is standing before anything else can be acknowledged, would be undermined by the existence of multiple universes, let alone of an infinite number of them. There is no way to fix our location absolutely—in either space or time. And even if there are not an infinite number of species that will come into being in the billion years still left to Earth before it incinerates, there is time enough for something to emerge from out of the gene pool that could very well be more stunning than, or at least as surprising as, our own humanity. One need not think of other planets or universes to confront squarely the doctrine of multiple worlds. Simply attending to the fossil record can accomplish this as well.

But at this moment, the taste of green-tea ice cream is more or less keeping me located here and now in space and time. And Andrea has returned from her perch on the bench with a gleam in her eyes. She has passed out several intricately folded origami to whoever drew near to observe her adept fingers in action.

January 7

Today, we have reached Nachi Falls—the end point of our pilgrimage. It is here that Kumi has arranged a meeting of the students and their professorial entourage with Asahi Guji, a celebrated personality and spiritual figure in the world of Shintoism. He serves as the head priest of the Shinto shrine at Nachi Falls, a home not only to many gods and goddesses but also to a stone that is, it is storied, the metamorphosed body of the three-legged raven who guided the first emperor of Japan across the Kii Peninsula to his future capital in Yamato Province.

Asahi addresses the class as Kumi again translates. I take down some all-too-sketchy notes of his words:

> Kumano means "from the depths." Japanese philosophy begins with the thought that we come to ourselves through others. We are born from our parents. Our parents from their parents. And so on. The depths of our beginning cannot be discerned. But this landscape upon which we stand plunges down toward it. The land instructs us: Our lives call for our gratitude.

Asahi continues by speaking of the recent cyclone, which left the shrine buried deep in mud and massive boulders. Only in the past weeks have buildings been unearthed, and repairs are now under way. The intensity and number of the landslides, Asahi notices, had much to do with silviculture practices introduced in the nineteenth century throughout the Kii Mountains and still in force. The forest was submitted to massive plantings of cedar and cypress, while all other flora were suppressed. This, I remind myself, was the final chapter in Ōkami's story as well, as the wolf was increasingly displaced by forests transformed into plantations (Walker 2005:35). Asahi implores us to strive to return these mountains to their natural conditions. Only then will they be able to receive the seasonal onslaught of cyclones with the resilience of which the land is truly capable. Disasters will, of course, still occur, but our hand in them should be kept to a minimum.

Afterward, Asahi accompanies us to the viewing site of Nachi Falls, where he offers us its waters to drink. "One swallow," Kumi translates for him, "and you will live a long life. Two swallows, and you will live eternally. Three swallows, and you will live fully until the moment you die." Mortality is the very vocation of our humanity, Asahi reminds us. All else is dross.

> Prayer sticks of pine burn
> to cinders: Only then can
> their ashes speak.

Later, at Yunomine, the site of an *onsen* (hot spring) and ancient spa, we are invited by a Buddhist monk into a small temple with an altar of intricately worked gold surrounded by a host of statues with food offerings set out before them. To one side of the altar, we find ourselves standing face to face with a black Buddha, who grimaces like a demon as he holds a double-edged sword upright in his right fist and offers a coiled cord to the viewer with his extended left hand. Louise, artist and fellow teacher, asks: "Isn't he rather warlike? Not what I readily think of as Buddha." I agree. "Only to our desires," Jasu answers. A teacher and volunteer trail ranger who has been walking with us on the pilgrimage, Jasu quickly earned our respect for his depth of insight and knowledge of all things Kumano Kodō. He continues. "This instantiation of Buddha slays the 108 forms of desire that we humans cling to. As long as we cling to these desires, Buddha remains ferocious in our eyes. And this is not a bad thing for us." Reading later, I discover that the sword is termed *prajña* (discriminating insight) in Sanskrit. Cutting both ways, the sword signifies the sharpening of perception and inquisitiveness, as well as the letting go of attachment to one's perceptions and inquisitiveness (Lief 2002; Low 2006:18). No place left to stand.

A Sermon:

We humans of the Anthropocene, so confident of where we are standing, cling to our manifold desires, even as they cost us the presence of species after species with whom humanity might have more generously shared Earth. The emergence over centuries of a global economy celebrating the multiplication of human desires and their autonomous satisfaction has been profoundly misleading. We have been deluded into thinking that we are in control. But Ōkami and the supermajority of the extinct provide, in their very state of having once existed in flesh and blood, telling testimony that this is not the case, for us or any other species under the sun. We of the global first world are in need of a warrior Buddha with the strength and ferocity to confront our spectacular inattentiveness and our misplaced fondness for specimens and lists, for observing from afar the mayhem we have inflicted on the face of Earth in a time of mass species extinction. The warrior Buddha would awaken each of us to all the manner of misdirection at work in the technocratic imperium dominating our time and place.

And might not this task fall as well to Ōkami and all the other species who have been rendered ghosts in our time? Even if they no longer are instantiated as flesh and blood in the womb realm, might they not return from the diamond realm to repopulate Earth, each in its specific instantiation of the black warrior Buddha?

Notes for an Encylopedia Entry:

The Honshu wolf, known as Ōkami in Japanese, was also called *yama-inu* (mountain dog). Smaller than other wolves, Ōkami fed on rodents and other small animals but also effectively preyed on *shika* (a small elk) and *inoshishi* (boar). Ōkami was also known to have persistently disturbed the graves of the newly dead, feasting on their insentient flesh. All of them now rest easier with Ōkami's passing. Today *shika* and *inoshishi*, no longer kept in check by the hunger of Ōkami, overrun forest and field (Côté et al. 2004; Takatsuki 2009). The extinction of one species, as mountain after overgrazed mountain testifies, has led to the overabundance of others. For generation upon generation, stretching back as far as word of mouth can report, Ōkami was considered an economic blessing to rural peoples and a *Gokenzoku-sama*, an honored intercessor for the gods. Beloved by humans, Ōkami dissappeared nonetheless, due to a change in farming practices, as well as to the sudden and devastating appearance of rabies in Japan. That Ōkami can have gone extinct in our time gives witness to how fragile the hold of manifold species on their place under the sun has turned out to be.

JANUARY 9

This evening, at Cherry Tree Abbey on Mount Koyosan, the Shingon monk serving us dinner remarked that twenty minutes of sitting meditation is just enough. More than that is dangerous, he added, since many abysses lie in wait in one's mind. Those who plunge into the practice of meditation are in danger of not returning. Those who step carefully will find that in the sitting itself instruction is given.

I wonder whether meditating on the abyss in time hollowed out by the extinction of the Honshu wolf for more than twenty minutes daily may run dangers similar to those the monk has mentioned. If so, then the return of Ōkami, however ghostly, from time to time to accompany one's walking the Kumano Kodō may be a phenomenon well worth attending to. And there is a precedent. At Tamaki Shrine, located in the heights of the Kii Mountains, the wolf played the role of an intecessor between the gods and humans in the curing of fox sickness (Knight 2006:207). When asked about this later, Kumi responds: "The fox is regarded as having magical power, and fox possession, *Kitsune-tsuki* [overcome by fox], is a kind of 'madness' that is normally temporary. You may be disoriented, disillusioned or confused."

In the face of irretrievable loss, of abysses in time and space that cannot be negotiated straightforwardly by the human mind or heart, no matter how knowing, no matter how conceptually adept, no matter how compassionate it might be, a wolf *Gokenzoku-sama* of the extinct might prove helpful. The project is impossible yet necessary. And so this journal. And so Ōkami, with fangs that cut both ways.

I think of the three venerable Buddhist nuns whom Kumi has visited in Kayoiura on Omijima Island, who say daily prayers for whales harvested and slaughtered in centuries past by the fishermen of their village. The whales are remembered among the departed—both human and more than human (Kato 2007:302). What was begun in Kayoiura might find a new relevance on the Kumano Kodō, as one offers prayers to commemorate as well an entire species. The question of what ceremony or liturgy might be offered for those species that have recently become extinct among us is not a light one. Rather than entombing the memory of the extinct as we gather the pelts, the bones, the recordings, the photographs, the accounts of their habits and demise, and then store and catalog them in museums and archives, might we not instead call out to Ōkami and all other departed species, imploring them to return? If not in flesh and blood, then at the very least in dream and vision? John Knight (2006:196) recounts the all-night vigils of *sasoidashi* (luring out) performed by wolf enthusiasts and naturalists in the Kii Mountains. Recordings of howling wolves are broadcast in remote locations in order to elicit the howls of real wolves in response. Perhaps the obsessions at play in the stratagems of a new breed of crypto-zoologist, reluctant to accept the reality of Ōkami's extinction, are symptomatic of a regrettable inability to mourn the disaster that has come upon these mountains with the loss of a remarkable living being. But the commitment of the "wolf callers," as Knight terms them, to the relevance of Ōkami is worthy of respect. Better that those recordings be played for as long as humans inhabit these forests than that Ōkami's name be dispatched to an archive. Perhaps the ritual that the extinction of Ōkami calls for has already begun to be performed.

CODA

Oguchi no Magami, Large-Mouthed and Pure God,
    in the womb realm
of compassion, you were the bearer of swift death. Jaws gaping
and muzzle bloodied, you preached mayhem to the
    ravens, softness
to the waters, indifference to a falling star.

Recently marooned in the diamond realm of acute
    mindfulness,
you are giving birth to and suckling an infinite congregation
of four-legged, hairy Buddhas from whose inopportune
    howling

forgotten elements are arising, and all suffering, it is
    promised,
is to begin anew . . .

NOTES

1. Kūkai, also known after his death as Kōbō Daishi, was a Japanese
monk and philosophical thinker of the ninth century of the common
era who founded the tantric tradition of Shingon Buddhism and located
its center at Koyasan, in the vicinity of the Kumano Kodo. Among his
contributions to Buddhism was his insight into the Buddha nature of
all entities, both sentient and insentient.

2. Unless otherwise attributed, haiku in this journal were composed
by its author.

3. Free rendering of the original haiku based on Gabi Greve's (2006)
translation.

4. Free rendering of the original haiku based on Greve's (2006)
translation.

REFERENCES

Abram, David. 2010. *Becoming Animal: An Earthly Cosmology*. New York:
    Pantheon.
Barrow, Mark. 2009. *Nature's Ghosts: Confronting Extinction from the Age of
    Jefferson to the Age of Ecology*. Chicago: University of Chicago Press.
Côté, Steeve D., Thomas P. Rooney, Jean-Pierre Tremblay, Christian
    Dussault, and Donald M. Waller. 2004. "Ecological Impacts of Deer
    Overabundance." *Annual Review of Ecology, Evolution, and Systematics*
    35:113–147. doi:10.1146/annurev.ecolsys.35.021103.105725.

Greve, Gabi. 2006. "Wolf, Japanese Wolf (ookami)." *World Kigo Database*. http://worldkigo2005.blogspot.com/2006/11/wolf-ookami.html.

Hogan, Linda. 1995. "A Different Yield." In *Dwellings: A Spiritual History of the Living World*, 47–62. New York: Norton.

Kato, Kumi. 2007. "Prayers for the Whales: Spirituality and Ethics of a Former Whaling Community—Intangible Cultural Heritage for Sustainability." *International Journal of Cultural Property* 14:283–313. doi:10.1017/S0940739107070191.

Kemarre Turner, Margaret. 2010. *Iwenhe Tyerrtye: What It Means to Be an Aboriginal Person*. Darwin: IAD Press.

Knight, John. 2006. *Waiting for Wolves in Japan: An Anthropological Study of People–Wildlife Relations*. Honolulu: University of Hawai'i Press.

Kūkai (Kōbō Daishi). 2004. "The Meaning of the Word Hum." In *Shingon Texts (BDK English Tripitaka)*, edited by Mayeda Sengaku et al.; translated by Rolf W. Giebel. Berkeley, Calif.: Numata Center for Buddhist Translation and Research.

Lief, Judy. 2002. "The Sharp Sword of Prajna." *Shambhala Sun*, May 2002, http://www.lionsroar.com/the-sharp-sword-of-prajna.

Low, Albert. 2006. *Hakuin on Kensho: The Four Ways of Knowing*. Boston: Shambhala Press.

Mathews, Freya. 2003. *For the Love of Matter: A Contemporary Pan-Psychism*. Albany: State University of New York Press.

Mooney, Edward. 2009. *Lost Intimacy in American Thought: Recovering Personal Philosophy from Thoreau to Cavell*. New York: Bloomsbury Academic.

Takatsuki, Seiki. 2009. "Effects of Sika Deer on Vegetation in Japan: A Review." *Biological Conservation* 142, no. 9:1922–1929. doi:10.1016/j.biocon.2009.02.011.

Walker, Brett. 2005. *The Lost Wolves of Japan*. Seattle: University of Washington Press.

Weston, Anthony. 1994. *Back to Earth: Tomorrow's Environmentalism*. Philadelphia: Temple University Press.

PLATE 28.

MIDAS ROSALIA.

Lizars sc.

"Midas Rosalia." (From Sir William Jardine, *The Natural History of Monkeys* [Edinburgh: Lizars, 1833]. National Library of Australia)

2. SAVING THE GOLDEN LION TAMARIN

MATTHEW CHRULEW

The *mico* perches on the open threshold of her wood-and-wire cage, listening out into the alien forest. Twin young hang from her shoulders. Her mate and more twins chatter nervously behind her. Strangely, the door has been opened by the admittedly strange creatures who brought them in boxes to this odd enclosure, watched and scribbled notes for some days, and then released them here amid the dense trees, no walls any longer keeping them captive— or safe.

Unfamiliar smells and sounds filter through the foliage. They hesitate before this disconcerting place, native made foreign. This *is mico* forest. Forever their kind were born here, raised here, lived here, died here. But mostly it is gone now—logged and cleared for building and ranching, fragmented and needing care. And so gone, too, are their wild cousins, so diminished that, without bolstering, they could soon be extinct. Yet this lot have themselves been gone so long that they have forgotten the forest and can no longer interpret its angles and depths or hear the spectral whispers of their kin who graced its boughs. Born amid simulated trees in the middle of cities, this poised troop does not know to long-call out, to set their bounds. They do not know where to find fruit or how to swing on thin branches or not to walk the forest floor. They have little inkling of the many difficulties and threats that await: insects and swamps, hunters and snakes, hunger and thirst. Their managers have their own ignorance, too—they have forgotten the *micos'* learning, misjudged their immured naivety. A wild stupidity.

The *mico* snatches a bug from the air and eyes the untried pathways ahead. Beyond lie new risks, new lives, an adventure, yet death and suffering, too. Their existence is now a threshold experiment. They sit on a border so heavily burdened for their keepers, between wild and captive, nature and culture. Throughout history, things have largely gone one way (hunted, domesticated, collected, rescued . . .). Yet today, with the golden blur of one youthful *mico's* courageous dash into the canopy, these agile primates cross it in reverse.

Golden lion tamarins (GLTs) are slender, nimble 1-pound monkeys with long prehensile tails and engaging chattering mannerisms, native to the now mostly cleared Mata Atlântica (Atlantic Forest) of southeastern Brazil. Scientifically designated as *Leontopithecus rosalia* yet known as *micos* in everyday Portuguese, these remarkable endangered New World monkeys—callitrichids like the marmosets—are, with their silky golden-red coat and leonine mane surrounding an expressive, hairless face, the best known of the lion tamarins.[1] Their small cooperative social groups occupy territories in lowland coastal forest of Rio de Janeiro state, in which they forage for fruit, bark, nectar, and insects and other small animals, sharing the responsibility to carry and care for the offspring of a sole childbearing female.

The once-vast tropical rain forest in which tamarins evolved—a rich ecosystem recognized as a "biodiversity hotspot" supporting numerous endemic species—has now largely been destroyed. As Warren Dean (1995) recounts, the logging and clearing of the Atlantic Forest, its reduction to "bare ground" (364), has meant the catastrophic decline and disappearance of numerous species. While an ongoing process since the arrival of Portuguese colonists in the sixteenth century, much of this deforestation has been remarkably recent, a result of Brazil's often unrestrained twentieth-century development in service of a rapidly growing population largely concentrated in this region. Forested land has been logged for lumber, converted to agriculture or cattle pasture, and cleared for rail, roads, or urban development. Only very limited relict patches remain that, even when protected, are still vulnerable to clearing, damming and drainage, natural and malicious fires, exotic species, and the hunting and capture of wildlife (Chiarello 2003). Conservationists have worked to resist and reverse this deforestation—to preserve and protect habitat, to reconnect remnants by planting corridors, to change laws and policies and public and corporate attitudes and behavior—but have struggled against a mind-set that prioritizes economic development ("at any cost") over environmental concerns. Today, continued negotiations and inventive alliances are required to meet the genuine social justice demands of the Sem Terra landless agrarian reform movement at the same time as preserving what remains of the fragmented and degraded Atlantic Forest and its native threatened species (for example, Cullen, Alger, and Rambaldi 2005; Pádua 2013).

The waning of the golden lion tamarins has, of course, mirrored that of their forest. Since their discovery by colonists in the Renaissance, they had been exported to Europe as aristocratic pets. Still seen as common in the nineteenth century, they were subjected to intense hunting and trapping and other depredations. They were popular exhibits, "obviously most desirable as zoological-garden inhabitants" (Crandall 1964:101). In the postwar period, hundreds per year were taken from the wild, most often legally, for captive exhibition, biomedical research, and the pet trade (Mallinson 1996). Combined with drastic loss of habitat, for a species already restricted to small ranges of vulnerable lowland forest, these factors led to the dramatic decline of GLT numbers such that, by the mid-twentieth century, they rapidly approached extinction.

It was only in the 1960s, as a result of the work of leading Brazilian primatologist Adelmar Coimbra-Filho and others (Coimbra-Filho and Mittermeier 1977), that scientific, governmental, and public attention began to be drawn to their plight. Coimbra-Filho's surveys estimated that there were only a few hundred left in the wild. GLTs began to be officially recognized as "endangered," and (though not before a final hurried intake) zoos pledged to no longer import them. Scientists gathered in 1972 for a significant conference, "Saving the Lion Marmoset" (Bridgwater 1972), which would be followed by numerous others. In 1974, an area of lowland forest east of Rio de Janeiro was eventually protected specifically as GLT habitat—the Poço das Antas Biological Reserve, the first such sanctuary in Brazil. Much work was done in the following years and decades: prominent zoos (such as the Smithsonian's National Zoological Park in Washington, D.C.; the Frankfurt Zoo; and the Jersey [now Durrell] Wildlife Preservation Trust) became heavily involved; international committees for GLT conservation and management were formed and, eventually, localized in Brazilian hands; and funding from organizations such as the International Union for Conservation of Nature and the World Wildlife Fund was sought and pledged. Husbandry and management protocols were developed and refined; the viability of the GLT population and habitat was quantitatively analyzed and computer simulated; land and infrastructure were purchased; professional and community environmental-education programs were set up; and long-term studies of wild tamarins' behavior, demography, and socioecology were carried out in order to support their reproduction and survival.

Alongside these in situ preservation and research efforts, zoo biologists, primatologists, and others increasingly devoted their attention to GLTs' ex situ conservation—that is, managing the captive population as a genetic and demographic reservoir to insure against their extinction in the wild. In 1972, about seventy GLTs were kept in zoos throughout North America and Europe—a small, fragmented, and declining artificial population. Still largely ignorant regarding many aspects of GLT behavior and social structure, keepers faced numerous difficulties in the husbandry of these seemingly "too delicate" animals (Crandall 1964:101). Their health and longevity were frequently poor: they seldom reached the fifteen years they might in the wild, and they were often malnourished or inbred, suffering from rickets, "wasting marmoset syndrome," and diaphragm hernias, among other conditions. They harmed one another and themselves, and rarely bred in captivity or neglected their offspring when they did. But with importation banned by both Brazilian and American laws from the late 1960s, they could no longer be replenished from the wild. The viability of the captive population now relied on solving these problems. Zoo biologist Devra Kleiman established a captive management and research program for GLTs, investigating their social and reproductive behavior and developing guidelines for their care. Knowledge about the species' biological particularities and needs grew and fed into the refinement of husbandry protocols focused on nutrition, enclosure design, and particularly the formation of suitable family groupings in which, rather than being removed, juveniles were able to play a role in rearing infants so as to observe and learn parenting skills (Kleiman 1977; Kleiman, Ballou, and Evans 1982). Previously competing institutions came to cooperate under the direction of a studbook keeper and management committee overseeing each individual and the population as a whole—*omnes et singulatum*—in terms of genetic diversity and demographic stability (Ballou 1992). Soon, these zoos began to have some rather dramatic reproductive success. Indeed, in the 1980s, after an initial period of rapid population growth, this captive breeding program became so excessively successful that the problem for these arks was now one of exceeding their carrying capacity. But numbers in the wild remained critically low.

Thus it was that, in late 1984, the first group of captive-born golden lion tamarins was reintroduced into the Poço das Antas Biological Reserve in a desperate attempt to boost the wild-living population, which was in real danger of extinction. A small, specially chosen group was given survival-skills training at the National Zoo in Washington, quarantined and flown to Brazil, quickly acclimatized, and released into the forest under the telescopic lenses of media cameras and the radiotelemetric ears of the tamarins' nervous custodians. This was one of the first times such a thing had been attempted, a significant event in the history of biology and, it turned out, a truly revealing experiment—one that exceeded, in both interesting and tragic ways, the expectations and plans of those involved.

Results were at first very poor. Observing them prior to release, a local expert in wild *mico* behavior called the zoo-borns "plastic monkeys," like toys (Beck 2013:41), and this assessment was borne out in their fates. They couldn't orient themselves, didn't move well through the trees, and didn't really know how to find food or avoid danger. Released with minimal follow-up assistance, the first reintroductees had what is called a *high mortality rate*; that is to say, they died of snakebite and bee sting, exposure and starvation, often within just days of release. Accustomed to the heavily maintained zoo environment, they simply did not possess the knowledge and skills to survive.

Despite these initially disappointing outcomes, the team persevered over the following years, dealing with inherent and arising difficulties, adjusting their techniques each time—training and provisioning, pre- and postrelease—and taking careful data on the results of this experiment in wild life. Survival rates at first remained low—less than one-quarter survived the first decade— and the animals still regularly required assistance long after release.

Eventually, however, enough reproduced in the reserve, and their offspring found it easier to navigate the forested world they were born into. By the third generation, closer to three-quarters would survive. The preserve was enlarged, and neighboring private land-owners were encouraged to conserve their own forest remnants, into which further captive-born GLTs were introduced. Another sanctuary was created (União Biological Reserve, to the northeast), to which newfound isolated groups and those threatened by de-velopment were translocated. Ultimately, with the stable and "self-sustaining" captive population continuing to provide a sur-plus, the wild population itself stabilized, and the project to save the golden lion tamarin was deemed a success.

The story of the golden lion tamarin reintroduction is a famous and complex one that can be told in many different ways. As Rick De Vos (2007) points out, extinction itself as an event is difficult to know or experience; it is a matter of the articulation and performance of *dis*appearance and absence. Yet the storying of extinction—and of efforts to *counter* it, in particular—does tend to fall into familiar heroic narratives (Turner 2007). The project to save the GLT has been described and analyzed in academic arti-cles, feted in popular-science narratives, and even novelized; most often, it has been celebrated as the foremost example of a success-ful reintroduction of an endangered species. The scientists in-volved have been active in recounting its history and the role they or their institutions played in its successes (Kierulff et al. 2012; Kleiman and Rylands 2002). The zoo community that once con-tributed so heavily to the endangering market for wild tamarins "later became the salvation for these marvellous creatures" (Mallinson 1996:201; see also Kierulff et al. 2012:39; Mallinson 2003). Regular progress reports are provided in the newsletter *Tamarin Tales*, initially edited by the original GLT studbook keeper, Jonathan Ballou. Jeremy Mallinson's (2009:187–193) panegyric on Gerald Durrell boasts of the Jersey zoo's involvement in GLT research and conservation. Leading primatologist Benjamin Beck, who with Kleiman and others from the National Zoo was a central figure in the reintroductions from the start, has given a fictionalized account in his amateur novel, *Thirteen Gold Monkeys* (2013), a story of "hope, love, and unspeakable death in a disappearing Brazilian rainforest." While it expectedly lauds the determination and sacrifice of both the scientists and the tamarins involved, the book also does, for its part—amid a medley of often self-justifying, not to mention awkward and conflicted anthropo- and zoo-morphisms— attempt to explore the mistakes, challenges, and difficult and often heartbreaking decisions the scientists faced. Others have reported on the project from the outside, some more critical of its insufficiency (for example, Dean 1995:356–359), but most praising its success, with varying levels of enthusiasm for captive breeding and reintroduction as the future of conservation (for example, Luoma 1987; Tudge 1992).

The *micos* themselves have become celebrated icons. They are a charming and charismatic species, eliciting strong public interest and sympathy, that has been successfully promoted as a conservation symbol through media and educational campaigns (Dietz 1998). They feature on the new R$20 banknote, while the bid for them to be the mascot of the Rio Summer Olympics in 2016 was pipped by a cartoonish hybrid that nonetheless resembles them. Indeed, they have become almost synonymous with Brazil's natural heritage: a consummate "flagship species" using their famous and protected status to help mobilize support and funding for regional conservation and thereby, we are reminded, to also protect wider ecological systems and the habitat of other, less loveable species (Dietz, Dietz, and Nagagata 1994).

Today, the project is acclaimed as a pioneering exemplar of endangered species reintroduction and, indeed, as a model for multidisciplinary and international conservation initiatives that engage with the wider community and in which zoological gardens are prominently involved (Mittermeier 2002:xvii). If questions are asked about the "cost" of the project, it is only in terms of financial cost effectiveness and efficiency (Kleiman et al. 1991).[2] In addition to advancing biological knowledge and conservation techniques, it boasts of many wider achievements: successful lobbying of governments and fund-raising through zoos and NGOs, designing and implementing effective environmental-education and -training programs, and leveraging the profile thus generated to enroll previously resistant local ranchers and others as protectors of GLT habitat. It is often lauded as a cutting-edge, proactive practice, developed at the forefront of reintroduction science (Kleiman 1989), forging the path for counter-extinction and rewilding efforts, an expiation of zoos' past wrongs, and a ray of hope in a grim era of dwindling biodiversity.

Most of all, the species has been saved. The wild population grew and stabilized, to the extent of the preserved forest's capacity. Occasional heightened threats from predators, poachers, and competing "exotic" primates were identified and (as far as possible) mitigated. The appalling early death rates were brought under control; the scientists learned their lessons, and learned what lessons the zoo-born *micos* must receive in order to survive in the wild, what structure their own program must take to facilitate the GLTs' successful adaptation. Of course, the pragmatic limits of such projects are often recognized: the situation is still very delicate, the boosted numbers not enough to guarantee survival long term; and while the project has helped to protect and restore areas of forest, accruing further fragments and corridors, there is never enough new habitat to which GLTs can expand.[3] But the status of GLTs on the IUCN Red List of Threatened Species was downgraded from critically endangered to endangered in 2003, and the project has transitioned from reintroduction to maintenance and from American to Brazilian administration. According to the major criteria set out—survival and reproduction, a self-sustaining population (Kierulff et al. 2002:272; Kleiman et al. 1991:139; Kleiman, Stanley Price, and Beck 1994:288)—the project to save the golden lion tamarin was an indubitable, if expensive, success (Castro et al. 1998:125; Kierulff et al. 2012; Kleiman et al. 1991).

But is that the end of the story? Is that all there is to be learned? There are wider questions to be asked of this important, yet also often troubling, counter-extinction practice, different stories to be told (Rose and van Dooren 2012), and alternative versions to be cultivated (Despret 2016:169–176) that should enable us to open up new capacities for response. The situation is historically, politically, and culturally complex, and made more so by the cultural and behavioral complexity of the particular animals involved.[4] Without denying the pressing need to resist the horror of extinction— indeed, in service to this very ambition—we must ask *at what price* (and courting what dangers, tolerating what failures) are species rescued from this fate?[5] What costs were borne by these tamarins? Of what were they (made) capable—and incapable? Can we tell the story in a way that attends more widely to the multiplicity of these costs and achievements, and that problematizes them in turn? That distinguishes the multiple registers of human intervention and animal (in)ability being judged and tested in this life-and-death experiment? Can we navigate the complexity of human–animal entwining in processes of extinction *and* attempts to counter them?

The fates of the first reintroductees were unexpectedly harsh. Their deaths occurred remarkably quickly, in the first few days after release, often in violent and painful ways. We should not move on too quickly from the brutality of this fact. Their "sacrifices"— necessary and unnecessary—deserve to be lamented. Yet to do so, we also need to understand the ways in which these lives and deaths were articulated by the biopolitical dispositive that produced them. The scientific articles recount this mortality quantitatively and in objective language. In Beck's novel, the events are described with more pathos. The tamarins' struggles to find food and safe sleeping sites, their efforts to survive the attacks of both predators and competitors, and their anguish at the theft of their young are used as sources of narrative tension, and provoke the characters' emotional and pragmatic reflections. Having initially given the first GLT group heroic names from Greek mythology, the biologists decide to refer to them by numbers in order to avoid too much emotional attachment (Beck 2013:107, 157, 231). Yet this detachment does not last long; they are inextricably and intimately bound to these golden monkeys.

Perhaps the most striking expression of the fortunes of the tamarins comes in the glossary at the end of the novel; at the same time as this cast of characters gives the names of the members of the Olympia Family released into the reserve's quarry area on December 7, 1984, it coolly lists their deaths alongside: Mom "disappeared on 8 December"; Dad was "eaten by a boa constrictor on 9 December"; of the twelve-month-old twins, Hera was "killed by bees on 11 December," while Hercules "disappeared in a group encounter on 9 December, [and was] found dead of starvation on 12 December" (Beck 2013:244–245). Only Pandora and Prometheus, the aptly named eighteenth-month-old twins, managed to survive longer than the first few days (by pairing up with others).

Such fatalities were hardly confined to this first group. Survival rates remained low over the first decade of the project, with less than one-quarter of the 136 reintroductees surviving until 1994. Captive-born tamarins died much more frequently in their first year than would otherwise be normal (Kierulff et al. 2002:276). With the GLTs dying quicker than they could reproduce, it took a dozen years for the reintroduced population to stabilize (Kierulff et al. 2002:277). The GLTs were exposed to a range of new hazards they found difficult to negotiate: "The major causes of loss (i.e., death, disappearance, and rescues) include theft by humans,

starvation, exposure, bee stings, disease, wounds resulting from social conflict, consumption of toxic fruits, and snakebite" (Castro
et al. 1998: 115–116).[6] They had to deal with parasites, biting ants
and bugs, and competition for food, territory, or mates. They
contracted viruses; were injured in encounters with predators, rivals,
or would-be poachers; or just regularly found themselves disoriented, cold, and hungry.

These fatal liberations constitute a considerably novel event in
the history of zoo biology. While zoos may have styled themselves
as protective arks, few initially believed that they would actually
achieve disembarkment into a recovered wilderness; in the 1970s,
Kleiman had not even thought reintroduction feasible (Kleiman
and Rylands 2002:xxii). Zoos had nonetheless long maintained a
belief in their animals' *wildness*: they saw their task as the preservation of natural behavioral expression. This demand, however, was
strongly (and often awkwardly) combined with the imperative to
actively *care* for their wards: to foster life; to nurture their health,
welfare, and happiness; to protect them from stress, disease, starvation, and predation—if not risk itself. Death, then, was something
that must be thoroughly investigated and, as far as possible, elimi-
nated from this biopolitical Eden (Chrulew 2013). This had
certainly been the case for GLTs, whose nutritional, behavioral,
social, and reproductive needs had been researched in detail so as
to optimize the captive population. So when, in 1984, in response
to the threat of extinction, Kleiman, Beck, and others trans-
ported captive tamarins from their urban Washington enclosures
to Brazilian lowland forest and released them there to see and to
know how well they would survive, the balance between zoo biology's conflicting demands of care and naturalism shifted toward a harsher settlement.

Exposing endangered animals to such palpable risks was unprecedented and at odds with every standard of good zookeeping. As
one of the tamarins reflects in Beck's (2013) novel, "This is no zoo"

(10). While the biologists had suspected that the *micos* might not quite be up to forest life, they had not expected such heavy losses. Though gender and other temperamental and professional differences qualified their responses, they were shocked to watch on as their precious wards died so horribly and so soon. In the zoo, they were the objects of extreme levels of care and investment.[7] All the ways in which captivity contributed to behavioral abnormality and mortality were carefully identified and removed.[8] Yet here they were, releasing the tamarins into a lethal environment for which they were clearly unprepared. When Hera, the favorite of Beck's (2013) protagonist, died of bee stings, "He held his emotion back, but said: 'Well, that's four lost out of eight, all in only five days. Not a great outcome'" (105). Indeed.

What went wrong? Such a result was always in the cards. Reintroduction is problematic at the best of times, and techniques were then still in their nascency (Kleiman, Stanley Price, and Beck 1994:295). But the critical status of the wild population called for desperate measures. What else could the biologists have done? They did have in their repertoire practices that might have curtailed such fatal results, but they opted for a "hard release," providing minimal postrelease support. Why were they happy to err on the side of risk rather than caution in expelling these monkeys from their Eden? No real benchmarks yet existed to guide them, and money and time also came in to it: Beck (2013:57–59, 181) dramatizes a heated quarrel, to which Kleiman (1989:158; see also Kleiman, Stanley Price, and Beck 1994:294) also obliquely refers, in which, due to limited resources, a decision was made to speed up the reintroduction and leave the GLTs to fend for themselves. Ultimately, such a severe decision could be made only because the reintroduced GLTs were classed as *surplus* to a thriving captive breeding program—what Vinciane Despret (2015) might call an *excess of achievement*. Were these groups, then, just "bare life" exposed to death in a state of emergency, individuals whose welfare and indeed lives were sacrificed for the salvation of the species (Beck 1995:157; Kleiman 1989:158)?[9]

Some might defend such a release according to a certain hard-nosed naturalism: Isn't suffering just a fact of life in the wild? Doesn't any real wildness inevitably include genuine exposure to the possibility of predation, starvation, and disease? Surely, these "wild" animals must at some point have to learn to cope with a natural environment, however unforgiving (on such questions, see Beck 1995)? But, despite some negotiations and compromises, this was not at all the scientists' position. Such outcomes were unacceptable and undermined the entire purpose of the reintroduction. These were animals who were personally known and cared for, for whom the keepers had acquired a real responsibility. Certainly, they would come to support some opening up of the system of care to contingency and risk. But the cruel outcomes of the initial hard releases were unacceptable in welfare terms. Moreover, they were not "natural" at all but at least indirectly anthropogenic, occurring only due to the reintroduction and often the result of the *micos'* history in captivity, their adaptation to the zoo or inadequate preparation for living in the forest—deathly artifacts in need of remediation. Thus while the first groups may have experienced a particularly arduous homecoming, in the end the overall program rejected any neo-Darwinian suspension of the biopolitical welfarist regime. Far from simply leaving the GLTs to the red claws of "nature," the scientists did all they could to help them survive, implementing a range of techniques designed to assist their transition to wild life.

Their first step was rehabilitation or "pre-release enrichment"—that is, submitting the GLTs to survival training in the zoo as preparation for reintroduction (for example, Box 1991; Castro et al. 1998; Kleiman 1989; Kleiman et al. 1986; Shepherdson 1994; Stoinski et al. 1997). This had already been attempted with the Olympians, though, as they discovered, inadequately so. So they redoubled their efforts, installing a number of new conditions designed to foster the skill, stamina, and agility needed to survive in the wild. Tamarins were made to search for their own food, to find where it was scattered or hidden, and to manage whole rather than cut-up pieces, breaking the food's connection with people in the process. They were exposed to "predators"—at least mimicked or restrained—so that they might learn to avoid (rather than trying to eat) poisonous snakes and toads. They were encouraged to use smaller, higher branches and vines and other, more realistic foliage and climbing apparatuses that were regularly shifted to disrupt habitual routes. They were left to range freely and deal with the weather and other hazards in larger, open wooded areas of the zoo, a "boot camp" free from their physical enclosure but not their "psychological cage" (Beck 2013:144–153; Beck and Castro 1994). Various other unexpected and challenging events were introduced into the routine at participating "gateway zoos" (Stoinski et al. 1997), both passively exposing the GLTs to more complex environments that obliged them to work and learn, to cope and adapt, and actively intervening to promote behavioral change and specific physical and social skills, such as foraging, navigation and locomotion, predator avoidance and defense, and interacting and communicating with their own and other species. This training regimen and enriched milieu thus attempted to lure these comfortable citizens of the zoo into becoming wild.

"Enrichment" was not a new thing in zoo biology; the introduction of stimulation, novelty, and certain "naturalistic" conditions into often sterile and monotonous captive environments had been used for decades as a way to counteract stereotypy and other abnormal behaviors and to improve animals' well-being (Gibbons et al. 1994 [particularly, Novak et al. 1994; Snowdon 1994]; Shepherdson 1994; Shepherdson, Mellen, and Hutchins 1998). It had become recognized as crucial, especially in the captive propagation of endangered species, to design environments that catered to animals' species-specific needs and well-being and that encouraged the retention of the full diversity of behavior (Hediger 1964). Important elements of the natural habitat were transposed, and attention was paid to relevant "behavioral indicators" (such as breeding success and normal behavioral expression) that signaled the quality of the environment.

What *was* new in the GLTs' reintroduction training, however, was the level of exposure to danger that was countenanced. Risk-averse zoos were hesitant to consider jeopardizing their wards. But to suitably prepare GLTs for reintroduction, zoo biologists found themselves needing to simulate precisely those aspects of nature that they had often boasted of removing: "[E]ffective preparation of captive primates for reintroduction requires environments that are more naturalistic—that is, that provide more stress and less well being" (Beck and Castro 1994:269–70; see also Beck 1995:155; Castro et al. 1998:124; Shepherdson 1994:172–173). Such proposals put the lie to supposedly "naturalistic" exhibits aimed more at visitor immersion (Beck 1995; Beck and Castro 1994), instead favoring the functional simulation of natural hazards and posing the question: "Are zoos willing and able to place animals at risk?" (Kleiman et al. 1991:141, 125). Certain types of stressful experiences, it was argued, are in fact healthy and conducive to well-being, particularly if the animals are able to respond appropriately and adaptively (Shepherdson 1994). Public-relations managers or ethics committees might protest, but at least in the case of threatened species like the GLT, the conservationist imperative overrode their concerns. The biopolitical care that sought to eliminate all danger gave way, however slightly, to a targeted simulation within the zoo of the natural difficulties and necessities of living in the wild.

Yet for all the zoos' efforts, it was ultimately determined that these enrichment interventions did not have the desired result. Beck and others compared the effects of the various environments, and found no evidence that pre-release training improved survival in the wild. Against the rarely tested expectation that training would enhance the tamarins' prospects on release, the data from their retrospective analysis of the success of different groups of re-introduced animals told them that GLTs exposed to free-range conditions or trained in survival skills prior to release fared no better than those not trained or those trained after release (Beck et al. 2002:294; see also Beck 1995:161; Beck and Castro 1994; Box 1991; Castro et al. 1998; Kleiman 1989). Any results they saw didn't persist in the wild and sometimes not even in captivity. While enrichment might promote some natural behaviors—free-range conditions more effectively than zoo training—both approaches were ultimately inadequate (too compromised, not long enough or during the critical learning periods, or just not realistic or challenging enough) to produce the necessary range of abilities.

Faced with this outcome, the focus turned to a more interventionist model of management, training, and support *after* release. Rather than hoping that the *micos* would retain or learn enough skills to survive a hard release, the scientists did all they could to ensure that survival through intensive "post-release provisioning"—what is called "soft release" (Beck et al. 2002; Beck et al. 1994:271). While the GLTs were still in their "psychological cage" (Beck and Castro 1994)—that is, not venturing "freely" into the forest but returning to the familiarity of their nest—the biologists fed and located them and otherwise gave them aid. They provided them with shelter boxes from which to begin their explorations, and put out food and water at progressively farther intervals. If the *micos* got lost, they brought them back to their group or nest; if they got hurt or sick, they fixed them up. Where possible, they paired them with wild-born tamarins they could learn from and kept them away from others that might compete with them for food or territory. They continued their practices of trapping and visual and telemetric monitoring, which increased their capacity to detect problems and intervene if necessary. In addition to improving the welfare of the animals, such assistance gave them enough time to gradually adjust, making them "more likely to survive to experience their new environment, to learn adequate responses to new situations, and to reproduce" (Castro et al. 1998:125). While some worried that this would only slow their progress, such slower, softer rewilding seemed much more humane, realistic, and necessary than the hard-and-fast fall of the first attempts (Beck 2013:133–135).

It is this practice of intensive postrelease management—and its progressive reduction—to which the success of the project is largely attributed. It reduced the number of untimely losses, allowing the zoo-born *micos*, who otherwise would have struggled, to survive for longer in the forest. They dispersed, joined up with wild-born *micos*, and formed new groups. They learned new ways of being. Most importantly, they reproduced. And their wild-born offspring, despite the behavioral deficiencies of their captive-born parents, were able to learn more quickly and negotiate the forest more easily (Beck et al. 2002:299; Kierulff et al. 2002:276; Kleiman et al. 1991:140; Stoinski et al. 2003). The goal of the reintroduction ultimately became less about the initial GLTs released than about the arrival and survival of their more capable and adaptable progeny: "Indeed, we now consider that the re-introduced zoo-born golden lion tamarins are living in a 'wild zoo' and that their purpose is to reproduce and provide offspring that will be truly self-sustaining" (Kleiman et al.1991:140; on "wild zoos," see also Marsh 2003).[10] By testing and adapting their techniques, then, after early botched efforts, the scientists contrived the most productive methodology to keep at bay the impending extinction of this species: securing one dependent generation to ensure the independence of the next.

In the process, technologies of care and management honed in that distinctive heterotopia of human–animal contact made their way into the heterotopias of wilderness that have long served as their reflection and aspiration. The reintroduction of an endangered species like the GLT provoked, alongside the internal opening up of zoo securitization to free-range enrichment and the simulation of stress and risk, the external *migration of the techniques of zoo biology* outside urban enclosures to new geographical zones of application: "They improve reintroduction success and the welfare of individual reintroductees. The strategies create a vital demand for zoo personnel, who already have the skills required for these sorts of postrelease management techniques. This, in turn, provides an essential link between the zoo, animal welfare, and in situ conservation and wildlife management communities" (Beck 1995:161). The end result of this growing integration of in situ and ex situ techniques is "metapopulation management," a centralized protocol overseeing the archipelago of fragmented forest and far-flung zoos, treating the wild and captive populations as a single interdependent unit between which transfers of animals and genes, as well as expertise and techniques, become ever more frequent and necessary (Kleiman and Rylands 2002; Marsh 2003).

Through the GLT reintroduction, zoo biological techniques of captive animal care were themselves introduced into the wild. They were at the same time transformed, opened up to the agency and subjectivity of their charges, whose conduct they needed to become able to conduct (Foucault 2002). Confronted with behavioral ecology's candid revelation that their animals were *plastic*, the zoo biologists, like their GLTs, had to adapt to their new environment. But they were prepared for the challenge: their entire apparatus had been constructed over decades as an ensemble of techniques for the production of wildness. And fragmented reserves like Poço das Antas offered the perfect opportunity to experiment: "[T]hey provide flexible and experimental reserves to better refine conservation management techniques" (Marsh 2003:370). Habitat fragments and species remnants thus formed a laboratory for the invention of interspecies technologies of power that sought simultaneously to intensively intervene and to erase their own impacts (Benson 2010; Reinert 2013; van Dooren 2016). This migration of animals, people, technologies, and techniques led also to the production of a new hybrid community, an experiment in shared life composed of multivalent interfaces between species (Lestel 2007).[11]

This experimental aspect of the reintroduction is crucial. It lacked the structure and refinement of the lab or the zoo, lurching headlong into the wild with its salvific urgency and cutting-edge naivety. But it was a remarkable experiment about learning and living, one attempting to *help GLTs survive* and at the same time to *work out how best to help them survive*. It was an empirical experiment that sought to learn not only about the tamarins themselves, but also about the scientists' own methodologies. They deployed and compared a number of release and preparation techniques, analyzing which ones worked best and recursively learning about learning and feeding back on their own operation as they went. It was simultaneously an ontological and a biopolitical experiment that sought to *make the GLTs live*, to secure their survival. They made knowledge by making tamarin history, producing the conditions of their objects' existence, as Isabelle Stengers (2000:146) puts it, yet in the extreme sense that the very future of this fragile form of life was itself at stake. It was a true experiment in wild life, an attempt to create an unprecedented multispecies community, that invented and enacted new natures, cultures, animalities, and subjectivities—new futures—amid overwhelming and singular losses. And it was a *test* in which how well the tamarins would survive—or how quickly they would die—was precisely what needed to be observed and learned, in order to be more optimally produced.

The GLT experiment was certainly presented with strong early feedback. For obvious welfare reasons, such starkly lethal results are inaccessible to more ordinary experiments. But here, in emergency conditions, the first reintroduced tamarins' untimely deaths provided quite revealing *data* from which much could be divined. After a steep initial learning curve, and its painful and lethal costs, the scientists developed better practices, ultimately settling on a more successful model. But they also had their expectations and presumptions confounded by new facts about the tamarins themselves. What was it that they learned?

Most significantly, they learned about tamarin learning. What they discovered, as a result of the GLTs' unexpected mortality, was just how much they are *plastic* and *cultural* animals. Their ability to survive in the wild is, in important ways, not simply innate or hard-wired but *learned*. That is, the biologists learned that the skills, knowledge, and customs that GLTs evolved for forest life are passed between the generations through social interaction rather than simply genetically; and, importantly, that this cultural transmission is interrupted in crucial ways in captivity, impairing the suitability of zoo-born *micos* for reintroduction. This is repeated in similar ways in a number of publications: that *critical survival skills are learned* and that *captive-raised tamarins lack proficiency in these learned critical survival skills* (for example, Beck et al. 2002:294). That is, at least in terms of surviving in the wild, captive-born GLTs are behaviorally "deficient" or "incompetent."

The logic by which this finding is articulated is often curious, to say the least. In one short work, "Behavioural Deficiencies in Reintroduced Golden Lion Tamarins Are Clues to the Effects of Successful Adaptation to the Zoo Environment," the researchers reason from the fact that "60% are lost in the first post-release year" to the conclusion that "many survival-critical behaviours may be learned" (Beck et al. 1998:7). The data of the GLTs' deaths are here presented as the premise of an inductive inference about tamarin behavior. But why was this something that had to be tested or proved? Why was it necessary that tamarins suffer and die in unfamiliar surroundings in order to demonstrate that they had grown accustomed to being cared for in the zoo? Was it not obvious to these experts that captive primates would not be quite so adept at forest life as those born and nurtured there?

Had they bought in too strongly to the zoo ideology of captive animal "wildness"? To an innatist model that assumed the captive GLTs would still have inherited the instinctive ability to survive in the forest? Or was there a quasi-behaviorist overestimation of the biologists' own ability to train the tamarins to do so? As we saw, they soon found that their efforts at enrichment could not overcome the handicaps of captivity. But it is not as easy as diagnosing here a naïve neo-Cartesian projection that reduced the monkeys to little instinctive machines.[12] These were experienced, sophisticated primatologists: the longitudinal behavioral-ecology field studies in Brazil of Coimbra-Filho and others are to New World primates what those of Jane Goodall and Jeanne Altmann are to chimps and baboons; while on the zoo side, Beck (1980), for example, had published the first major review of tool use among animals, demonstrating a real openness to often unacknowledged cognitive and behavioral capacities, from intentionality and deception to the manufacture and use of tools.[13] The fact that they attempted to train the GLTs from the start shows that they assumed a certain level of apprenticeship was required and likely missing. Yet it is hard to avoid the judgment that the scientists initially overestimated the instinctive innateness or programmable mechanicity of the tamarins' survival behaviors (for example, Snowdon 1994:224–225). The failures of the first reintroductions revealed an unfortunate mismatch between zoological ideas about the "wildness" of their charges and the tamarins' real ability to survive in remnant Atlantic Forest. The zoo biological apparatus for the preservation and production of the full repertoire of natural behaviors was here put to a most stringent practical test. It failed, and the animals paid the price.

Yet the rewards for these sacrifices were multiple. New scientific knowledge about GLTs and about the scientists' methods helped to secure the tamarins themselves. In this learned fable, the vital and instructive struggles of the tamarins, their urgent and often fatal attempts to overcome their behavioral and cognitive "deficiencies" and "incompetences," are frequently paralleled by those of the scientists themselves. Lessons were learned about tamarins' learning, about their adaptability and agency, their nurturing nature, their fragile and transformable cultural transmission of hard-won species-specific survival knowledge and behavior. The scientists learned, to paraphrase Spinoza and his recent interpreters from Gilles Deleuze to Vinciane Despret, that these animals affect and are affected, but more importantly, *how* they are affected, in what specific ways. They learned not only that many important abilities (such as foraging, orientation, food selection, predator avoidance, and communication) are culturally transmitted, that they must be learned through experience in an appropriate social group and environment, but that so too must the very ability to culturally transmit, to teach and pass on such behaviors: parenting itself must be learned, by observing and helping others to mother and nurture (Coimbra-Filho and Mittermeier 1977:85). They also learned about the limits of GLT learning: what things are less learned (the recognition of aerial predators seems more instinctive than that of snakes), what can be learned from others (wild-born partners were helpful tutors, while zookeeper training proved less effective), when is the best time to learn (when young), and the ways in which learning had been disrupted in captivity. They learned—even if they rarely express it this way—that their "incompetent" *micos* were not simply denatured but *decultured* animals.[14] Yet if in captivity the GLTs had been *made* plastic, molded and hardened into static immobility, it means that they *are* plastic: mutable, responsive, capable of learning, possessing their own inherent capacities for change (and being changed).[15] And more so than the tamarins' "nature," it was this very plasticity (Malabou 2008), the possibilities and limitations of their plastic animality, that became the object of the scientists' knowledge and techniques, opening up an *ethopolitical domain of intervention* that problematized the tamarins' patterns of behavior, their conduct, indeed their very subjectivity (Chrulew 2016). In the end, this ethopolitical experiment taught us what GLTs are capable of doing and learning, and how they are made and make themselves capable and incapable of living.

The scientists also learned about themselves and their techniques. Indeed, what was most acutely problematized throughout the entire experiment were the historical and ongoing impacts on the tamarins of human agency—the "artifacts" of captivity and scientific intervention. Zoological gardens and intensive wildlife management such as in the "wild zoo" of Poço das Antas Biological Reserve have often been criticized as merely producing "biotic artefacts" (Lee 2005:97; see also Spotte 2006), perpetuating the artifactualization of nature. Yet there is no more discriminating hermeneutic of artifactuality to be found than in the experimental apparatus of these reintroduction scientists. Faced with their first groups' behavioral deficits, they devoted enormous energy to cataloging the differences between wild and captive tamarins—physiological and genetic but, most important, behavioral differences, in vocalization, locomotion, reproduction, rearing, sociality, and also in habituation to humans, which made them susceptible to poaching and other dangers (Castro et al. 1998:118; Kleiman and Rylands 2002; especially, Stoinski et al. 2003). They exhaustively evaluated the causes of death among reintroduced tamarins and identified the anthropogenic factors that could have contributed (Kierulff et al. 2012; see also Beck et al. 1991; Kierulff et al. 2002:277; Kleiman, Stanley Price, and Beck 1994:297).[16] They were thereby faced with powerful evidence of just what had been lost of the unique golden lion tamarin ethos, developed over millions of years in the Mata Atlântica, the diversity of behavioral expression that had been extinguished with the capture of their form of life. They clearly identified the cause of these deficiencies as the tamarins' history in captivity; that is, they were *artifacts of an-*

thropogenic environments, the product of dependence on human care and provisioning. Their training and support interventions were attempts to mitigate these captivity effects inhibiting the GLTs' survival. Yet they also recognized that such interventions had their own iatrogenic effects—from the stress caused by capture and handling to the physiological effects of anesthesia—and thus worked at devising and deploying noninvasive methods (from ingenious lures for weighing to measuring hormones from droppings rather than blood) and at lessening their frequency and intensity. The ensemble of zoo-biological and wildlife-management techniques such as feeding and trapping constituted, of course, a significant intervention into the tamarin reserve. But targeted as it was toward self-erasure and the production of wildness, and recognizing as it was forced to the helplessness that captivity had wrought in these creatures, as well as their own agency and adaptability—their own rewilding powers—it transformed itself into a set of procedures aimed at gradually reducing their dependence on their keepers, at dismantling their psychological cage, at facilitating the production of more capable and independent, truly *self*-sustaining animals—and thus staged its own withdrawal.

Death—that ecological force so necessary to life—takes many forms. In countering extinction, it can arrive harshly or gently, can be given to some so that others can live. The reintroduction of golden lion tamarins into the wild was also a reintroduction of wildness into the animals' own secured and dependent lives—and thus a transformation of their relationships to themselves and others, to their environment, to danger and risk, to their kin (past, present, and future), and to death. By disrupting their intergenerational cultural transmission of the forest *mico* way of life, the passing down of what had been learned through untold reproductive labor over millennia, captivity had also interrupted their dialogue with their dead, debilitated the nurturing example of their elders, and silenced the nourishing voice of their ancestors (Rose 2004, 2006). The future of the golden lion tamarins, their very conditions of existence and form of life, will forever be marked by the wounds of their history in captivity and entwined with their ongoing maintenance. But the communal and generational renewal of the tamarins' own adaptability and plasticity, their infancy and surrogacy, will allow them to reknit their torn relational fabric and to revive their cultivated wildness. Their extinction would mean the eternal silencing of ancient and recurring voices. For now, both the dead and the living continue to call out—and to listen.

NOTES

1. This group also includes their golden-headed, black, and (latterly discovered) black-faced cousins. They could all be considered to belong to the same species, but are not defined as such for political reasons: given the central role played by such taxonomic labels in environmental policy, "conservationists have continued to classify the forms as individual species to ensure their continued legal protection" (Rylands et al. 2002:5).

2. Every surviving animal is estimated to have cost US$22,000 (Kleiman et al. 1991:125). In particular, the intensive labor of training, monitoring, and support are expensive compared with translocation and other conservation methods.

3. Most assessments agree that reintroduction is only ever a short-term solution, rarely successful, too focused on single species, and requiring intensive and expensive management, a stopgap awaiting true reforestation (for example, Balmford, Mace, and Leader-Williams 1996; Gipps 1991; Kleiman 1989; Konstant and Mittermeier 1982; Norton et al. 1995 [particularly, Beck 1995; Hancocks 1995; Loftin 1995]; Olney, Mace, and Feistner 1994 [particularly, Beck et al. 1994; Kleiman, Stanley Price, and Beck 1994]; Snyder et al. 1996).

4. On the general question of the complexity of human–primate interactions, see, for example, Fuentes and Wolfe (2002); Haraway (1992); Lestel (1995); Strum and Fedigan (2000).

5. As Jeffrey Nealon (2008:18) reminds us, these are central questions of the genealogical method—to ask, as Michel Foucault does, "What the conditions of this emergence were, the price that was paid for it, so to speak, its effects on reality" (Florence 2000:460).

6. Benjamin Beck (1995) provides a further description of the price paid by these animals:

We have seen a newly reintroduced golden lion tamarin sit immobile and shivering high in a tree in a cold (10°C) rain for twelve hours. . . . [A] young male . . . was found dead, wedged head-down in a tree cavity; he was presumably feeding on insects and could not turn around or back out of the hole. A female was fatally bitten by a coral snake that she successfully ate. A young pair with an infant tried to enter their nest box, which had been taken over by Africanized bees; the female was found dead on the ground, and the infant disappeared; the male was found hideously swollen on the ground. . . . A pregnant female was attacked by an ocelot but escaped as our observer distracted the cat. . . . The tamarin nonetheless aborted during the night and died two days later despite treatment. Most disturbing to contemplate is the stress on eight reintroduced tamarins that have been stolen from the forest and smuggled into the incountry pet trade. . . . Cold rain, Africanized bees, coral snakes, and poachers, all threats to well-being endured by wild and reintroduced tamarins . . . are risks that never would have accrued had the tamarins remained in captivity. (158)

7. Of course, it is only the cherished wards that enjoy such attention; zoos have long dealt death to animals categorized as food, pests, surrogates, surplus, or otherwise. What differentiates reintroduction is the willing subjection of *protected* species, the valued subjects of care for whom others are normally sacrificed, to the risks of suffering and death.

Yet it is entirely routine in another sense: the sacrifice of individuals in order to protect the species.

8. Indeed, captive GLT keepers ought to be painstakingly aware of possible causes of death among captive tamarins and the anthropogenic factors behind them:

> Fractures and traumatic mutilations at the time of capture are unfortunately quite common and due mainly to the use of inadequate traps and cages and lion tamarin maltreatment during transport. Sometimes they result from fights between unfamiliar and stressed animals confined in small boxes and cages. . . . Personnel who handle the animals may be ill prepared to do so. . . . Captive lion tamarins may also injure and even kill one another when housed in inappropriate social groups, when groups are too close to one another, and even when housed in natural family groups. (Pissinatti, Montali, and Simon 2002:257)

Stress "due to inadequate management" (261) (from deficient enclosures to other unsatisfactory practices) can lead to social, psychological, or physiological problems, including alopecia, self-mutilation, pneumoenteritis, and dystocia. Overall, "many of the noninfectious conditions may also be products of ex situ conditions and may be rectified by proper management of these primates" (267). Compare Heini Hediger's analysis of "death due to behaviour" in Chrulew (2013).

9. Jonathan Ballou (1992) lists among the questions to be considered in a captive breeding "masterplan": "If it is to be bred, with whom, when and where? If not bred, is it to be held for future breeding or is it surplus to the needs of the population (i.e., has it already fulfilled its demographic and genetic obligations)" (269). Indeed, the animals best suited for reintroduction will likely also be those most valuable to the captive population, the integrity of which is not to be compromised; thus it is the least genetically valuable that will be selected: "It is likely that initial releases will result in heavy losses of individuals. This should be an expected outcome of the program. Therefore, individuals that are

genetically valuable to the captive population should not be used for initial reintroductions," but rather "individuals that are genetically 'redundant' in the captive population" (270). On surplus animals, see also Chrulew (2011).

10. While the disciplines of zoo biology and wildlife management remain obsessed by defining and managing the wild, they are increasingly coming to recognize and think through the breaking down of the captive/wild distinction through human involvement at various levels (for example, Beck et al. 1994:267; Beck and Castro 1994:259).

11. Such is the pharmacological dilemma faced by conservation biology in the Anthropocene: an inescapable tangle of knowledge and sometimes vicious, sometimes virtuous spirals of intervention by which the conditions of "wild life" are negotiated and produced, where unprecedented relationships and responsibilities are formed between humans and animals, demanding new questions and approaches (HARN Editorial Collective 2015).

12. On the diagnosis of such crypto-Cartesianisms in the history of ethology, see Chrulew (2014).

13. On the establishment of Brazilian primatology, and particularly the growth of knowledge about the central role of (reproductive) female callitrichids beyond oversimplifications regarding monogamous behavior, see Yamamoto and Alencar (2000).

14. Perhaps "culture" is the missing element in conservation in more than one sense: not only must conservation biology, often scientistic and managerial, learn to take account of human cultures (including diverse attitudes to the use of animals, questions of poverty, and the politics of indigenous peoples and decolonization) (Fuentes and Wolfe 2002; Hoage and Moran 1998), as well as the cultures of science (Haraway 1992), but it must also confront the problem of *animal cultures* (Grundmann et al. 2001; Lestel 2003; Lestel and Grundmann 1999).

15. Such plasticity is not reserved to human beings in opposition to animal fixity, as tradition suggests, but characterizes both human and

nonhuman animal life. Such animal plasticity is, nonetheless, species-specific and otherwise differentiated. This plays out prominently in re-introductions, which require different techniques for reptiles and mammals; for fish and primates; for grazers, foragers, and predators; for zoo-borns and wild-borns; for the young and the old; and so on (for example, Box 1991:118; Konstant and Mittermeier 1982:70).

16. According to Maria Cecilia M. Kierulff (2012) and her colleagues:

> The main single cause of loss in the reintroduced population was theft and vandalism (21%). Problems with adaptation to the new environment, readily noticeable after the release of captive-born animals (e.g. inability to find food, and problems with locomotion and orientation), likely caused the majority of losses if considered together (starvation 13%; lethargy/diarrhoea/anorexia/dehydration 10%; hypothermia/exposure 10%; injuries 3%). Other causes of death were natural predation (15%), wounds from social conflict (8%), a bee sting, eating toxic fruit, snakebite and haemorrhage following miscarriage (21%). (39–40; see also Beck et al. 1991; Kierulff et al. 2002:277; Kleiman, Stanley Price, and Beck 1994:297)

Causes of death were quantified and known, and those that didn't match the "norms" of wild death—particularly because they were directly or indirectly anthropogenic, such as behavioral deficiencies due to captivity—were identified and targeted for intervention.

REFERENCES

Ballou, Jonathan D. 1992. "Genetic and Demographic Considerations in Endangered Species Captive Breeding and Reintroduction Programs." In *Wildlife 2001: Populations*, edited by Dale R. McCullough and Reginald H. Barrett, 262–275. New York: Elsevier Science.

Balmford, Andrew, Georgina M. Mace, and N. Leader-Williams. 1996. "Designing the Ark: Setting Priorities for Captive Breeding." *Conservation Biology* 10, no. 3:719–727.

Beck, Benjamin B. 1980. *Animal Tool Behavior: The Use and Manufacture of Tools by Animals*. New York: Garland.

——. 1995. "Reintroduction, Zoos, Conservation, and Animal Welfare." In *Ethics on the Ark: Zoos, Animal Welfare, and Wildlife Conservation*, edited by Bryan G. Norton, Michael Hutchins, Elizabeth F. Stevens, and Terry L. Maple, 155–163. Washington, D.C.: Smithsonian Institution Press.

——. 2013. *Thirteen Gold Monkeys*. Denver: Outskirts Press.

Beck, Benjamin B., and Maria Inês Castro. 1994. "Environments for Endangered Primates." In *Naturalistic Environments in Captivity for Animal Behavior Research*, edited by Edward F. Gibbons Jr., Everett J. Wyers, Everett Waters, and Emil W. Menzel Jr., 259–270. Albany: State University of New York Press.

Beck, Benjamin B., Maria Inês Castro, Tara S. Stoinski, and Jonathan D. Ballou. 2002. "The Effects of Prerelease Environments and Postrelease Management on Survivorship in Reintroduced Golden Lion Tamarins." In *Lion Tamarins: Biology and Conservation*, edited by Devra G. Kleiman and Anthony B. Rylands, 283–300. Washington, D.C.: Smithsonian Institution Press.

Beck, Benjamin B., Devra G. Kleiman, Maria Inês Castro, Jonathan D. Ballou, and Tara S. Stoinski. 1998. "Behavioural Deficiencies in Reintroduced Golden Lion Tamarins Are Clues to the Effects of Successful Adaptation to the Zoo Environment." *Advances in Ethology* 33:7–8.

Beck, Benjamin B., Devra G. Kleiman, James M. Dietz, Ines Castro, Cibele Carvalho, Andreia Martins, and Beate Rettberg-Beck. 1991. "Losses and Reproduction in Reintroduced Golden Lion Tamarins, *Leontopithecus rosalia*." *Dodo: Journal of the Jersey Wildlife Preservation Trust* 27:50–61.

Beck, Benjamin B., L. G. Rapaport, M. R. Stanley Price, and A. C. Wilson. 1994. "Reintroduction of Captive-Born Animals." In *Creative Conservation: Interactive Management of Wild and Captive Animals*, edited by P. J. S. Olney, G. M. Mace, and A. T. C. Feistner, 265–286. Dordrecht: Springer.

Benson, Etienne. 2010. *Wired Wilderness: Technologies of Tracking and the Making of Modern Wildlife*. Baltimore: Johns Hopkins University Press.

Box, Hilary O. 1991. "Training for Life After Release: Simian Primates as Examples." In *Beyond Captive Breeding: Re-introducing Endangered Mammals to the Wild*, edited by J. H. W. Gipps, 111–123. Oxford: Clarendon Press.

Bridgwater, Donald D., ed. 1972. *Saving the Lion Marmoset: Proceedings of the Wild Animal Propagation Trust Golden Lion Marmoset Conference*. Wheeling, W.Va.: Wild Animal Propagation Trust.

Castro, Maria Inês, Benjamin B. Beck, Devra G. Kleiman, Carlos R. Ruiz-Miranda, and Alfred L. Rosenberger. 1998. "Environmental Enrichment in a Reintroduction Program for Golden Lion Tamarins (*Leontopithecus rosalia*)." In *Second Nature: Environmental Enrichment for Captive Animals*, edited by David J. Shepherdson, Jill D. Mellen, and Michael Hutchins, 113–128. Washington, D.C.: Smithsonian Institution Press.

Chiarello, Adriano G. 2003. "Primates of the Brazilian Atlantic Forest: The Influence of Forest Fragmentation on Survival." In *Primates in Fragments: Ecology and Conservation*, edited by Laura K. Marsh, 99–121. New York: Springer.

Chrulew, Matthew. 2011. "Managing Love and Death at the Zoo: The Biopolitics of Endangered Species Preservation." *Australian Humanities Review* 50:137–157.

——. 2013. "Preventing and Giving Death at the Zoo: Heini Hediger's 'Death Due to Behaviour.'" In *Animal Death*, edited by Jay Johnston and Fiona Probyn-Rapsey, 221–238. Sydney: Sydney University Press.

——. 2014. "The Philosophical Ethology of Dominique Lestel." *Angelaki: Journal of the Theoretical Humanities* 19, no. 3:17–44.

——. 2016. "Animals as Biopolitical Subjects." In *Foucault and Animals*, edited by Matthew Chrulew and Dinesh Wadiwel, 222–238. Leiden: Brill.

Coimbra-Filho, Adelmar F., and Russell A. Mittermeier. 1977. "Conservation of the Brazilian Lion Tamarins (*Leontopithecus rosalia*)." In *Primate Conservation*, ed. Prince Rainier III of Monaco and Geoffrey H. Bourne, 59–94. New York: Academic Press.

Crandall, Lee S. *The Management of Wild Mammals in Captivity*. Chicago: University of Chicago Press, 1964.

Cullen, Laury, Jr., Keith Alger, and Denise M. Rambaldi. 2005. "Land Reform and Biodiversity Conservation in Brazil in the 1990s: Conflict and the Articulation of Mutual Interests." *Conservation Biology* 19, no. 3:747–755.

Dean, Warren. 1995. *With Broadax and Firebrand: The Destruction of the Brazilian Atlantic Forest*. Berkeley: University of California Press.

Despret, Vinciane. 2015. "Beasts and Humans." *Angelaki: Journal of the Theoretical Humanities* 20, no. 2:105–109.

——. 2016. *What Would Animals Say If We Asked the Right Questions?* Translated by Brett Buchanan. Minneapolis: University of Minnesota Press.

De Vos, Rick. 2007. "Extinction Stories: Performing Absence(s)." In *Knowing Animals*, edited by Laurence Simmons and Philip Armstrong, 183–195. Leiden: Brill.

Dietz, J. M., Lou Ann Dietz, and E. Y. Nagagata. 1994. "The Effective Use of Flagship Species for Conservation of Biodiversity: The Example of Lion Tamarins in Brazil." In *Creative Conservation: Interactive Management of Wild and Captive Animals*, edited by P. J. S. Olney, G. M. Mace, and A. T. C. Feistner, 32–49. Dordrecht: Springer.

Dietz, Lou Ann. 1998. "Community Conservation Education Program for the Golden Lion Tamarin in Brazil: Building Support for Habitat Conservation." In *Culture: The Missing Element in Conservation and Development*, edited by R. J. Hoage and Katy Moran, 85–94. Dubuque, Iowa: Kendall/Hunt.

Florence, Maurice. 2000. "Foucault." In *Aesthetics, Method, and Epistemology*. Vol. 2 of *Essential Works of Foucault, 1954–1984*, edited by James D. Faubion, 459–463. London: Penguin.

Foucault, Michel. 2002. "The Subject and Power." In *Power*. Vol. 3 of *Essential Works of Foucault, 1954–1984*, edited by James D. Faubion, 326–348. London: Penguin.

Fuentes, Agustín, and Linda D. Wolfe. 2002. *Primates Face to Face: Conservation Implications of Human–Nonhuman Primate Interconnections*. Cambridge: Cambridge University Press.

Gibbons, Edward F., Jr., Everett J. Wyers, Everett Waters, and Emil W. Menzel Jr., eds. 1994. *Naturalistic Environments in Captivity for Animal Behavior Research*. Albany: State University of New York Press.

Gipps, J. H. W., ed. 1991. *Beyond Captive Breeding: Re-introducing Endangered Mammals to the Wild*. Oxford: Clarendon Press.

Grundmann, Emmanuelle, Dominique Lestel, Aschta N. Boestani, and Marie-Claude Bomsel. 2001. "Learning to Survive in the Forest: What Every Orangutan Should Know." In *The Apes: Challenges for the 21st Century*, edited by George Rabb, 300–304. Chicago: Chicago Zoological Society.

Hancocks, David. 1995. "An Introduction to Reintroduction." In *Ethics on the Ark: Zoos, Animal Welfare, and Wildlife Conservation*, edited by Bryan G. Norton, Michael Hutchins, Elizabeth F. Stevens, and Terry L. Maple, 181–183. Washington, D.C.: Smithsonian Institution Press.

Haraway, Donna. 1992. *Primate Visions: Gender, Race, and Nature in the World of Modern Science*. London: Verso.

Hediger, Heini. 1964. *Wild Animals in Captivity: An Outline of the Biology of Zoological Gardens*. Translated by G. Sircom. New York: Dover.

Hoage, R. J., and Katy Moran, eds. 1998. *Culture: The Missing Element in Conservation and Development*. Dubuque, Iowa: Kendall/Hunt.

Human Animal Research Network (HARN) Editorial Collective, ed. 2015. *Animals in the Anthropocene: Critical Perspectives on Non-Human Futures*. Sydney: Sydney University Press.

Kierulff, Maria Cecília M., Paula Procópio de Oliveira, Benjamin B. Beck, and Andréia Martins. 2002. "Reintroduction and Translocation as Conservation Tools for Golden Lion Tamarins." In *Lion Tamarins: Biology and Conservation*, edited by Devra G. Kleiman and Anthony B. Rylands, 271–282.Washington, D.C.: Smithsonian Institution Press.

Kierulff, Maria Cecília M., Carlos R. Ruiz-Miranda, Paula Procópio de Oliveira, Benjamin B. Beck, Andréia Martins, James M. Dietz, Denise M. Rambaldi, and A. J. Barker. 2012. "The Golden Lion Tamarin *Leontopithecus rosalia*: A Conservation Success Story." *International Zoo Yearbook* 46:36–45.

Kleiman, Devra G. 1977. "Progress and Problems in Lion Tamarin (*Leontopithecus rosalia rosalia*) Reproduction." *International Zoo Yearbook* 17:92–97.

——. 1989. "Reintroduction of Captive Mammals for Conservation: Guidelines for Reintroducing Endangered Species into the Wild." *Bioscience* 39, no. 3:152–161.

Kleiman, Devra G., Jonathan D. Ballou, and Ronald F. Evans. 1982. "An Analysis of Recent Reproductive Trends in Captive Golden Lion Tamarins (*Leontopithecus r. rosalia*) with Comments on Their Future Demographic Management." *International Zoo Yearbook* 22:94–101.

Kleiman, Devra G., Benjamin B. Beck, James M. Dietz, and Lou Ann Dietz. 1991. "Costs of a Re-introduction and Criteria for Success: Accounting and Accountability in the Golden Lion Tamarin Conservation Program." In *Beyond Captive Breeding: Re-introducing Endangered Mammals to the Wild*, edited by J. H. W. Gipps, 125–142. Oxford: Clarendon Press.

Kleiman, Devra G., Benjamin B. Beck, James M. Dietz, Lou Ann Dietz, Jonathan D. Ballou, and Adelmar F. Coimbra-Filho. 1986. "Conservation Program for the Golden Lion Tamarin: Captive Research and Management, Ecological Studies, Educational Strategies, and Reintroduction." In *Primates: The Road to Self-sustaining Populations*, edited by Kurt Benirschke, 959–979. New York: Springer.

Kleiman, Devra G., M. R. Stanley Price, and Benjamin B. Beck. 1994. "Criteria for Reintroductions." In *Creative Conservation: Interactive Management of Wild and Captive Animals*, edited by P. J. S. Olney, G. M. Mace, and A. T. C. Feistner, 287–303. Dordrecht: Springer.

Kleiman, Devra G., and Anthony B. Rylands, eds. 2002. *Lion Tamarins: Biology and Conservation*. Washington, D.C.: Smithsonian Institution Press.

Konstant, William R., and Russell A. Mittermeier. 1982. "Introduction, Reintroduction and Translocation of Neotropical Primates: Past Experiences and Future Possibilities." *International Zoo Yearbook* 22:69–77.

Lee, Keekok. 2005. *Zoos: A Philosophical Tour*. Basingstoke: Palgrave Macmillan.

Lestel, Dominique. 1995. *Paroles de singes: L'impossible dialogue homme/primate*. Paris: Découverte.

——. 2003. *Les origines animales de la culture*. Paris: Flammarion.

——. 2007. *L'animalité: Essai sur le statut de l'humain*. Paris: L'Herne.

Lestel, Dominique, and Emmanuelle Grundmann. 1999. "Tools, Techniques and Animals: The Role of Mediations of Actions in the Dynamics of Social Behaviours." Translated by Nora Scott. *Social Science Information* 38, no. 3:367–407.

Loftin, Robert. 1995. "Captive Breeding of Endangered Species." In *Ethics on the Ark: Zoos, Animal Welfare, and Wildlife Conservation*, edited by Bryan G. Norton, Michael Hutchins, Elizabeth F. Stevens, and Terry L. Maple, 164–180. Washington, D.C.: Smithsonian Institution Press.

Luoma, Jon R. 1987. *A Crowded Ark: The Role of Zoos in Wildlife Conservation*. Boston: Houghton Mifflin.

Malabou, Catherine. 2008. *What Should We Do with Our Brain?* Translated by Sebastian Rand. New York: Fordham University Press.

Mallinson, Jeremy J. C. 1996. "The History of Golden Lion Tamarin Management and Propagation Outside of Brazil and Current Management Practices." *Der Zoologische Garten* 66, no. 4:197–217.

——. 2003. "A Sustainable Future for Zoos and Their Role in Wildlife Conservation." *Human Dimensions of Wildlife* 8:59–63.

——. 2009. *The Touch of Durrell: A Passion for Animals*. Brighton: Book Guild.

Marsh, Laura K. 2003. "Wild Zoos: Conservation of Primates *in situ*." In *Primates in Fragments: Ecology and Conservation*, edited by Laura K. Marsh, 365–379. New York: Springer.

Mittermeier, Russell A. 2002. "Foreword." In *Lion Tamarins: Biology and Conservation*, edited by Devra G. Kleiman and Anthony B. Rylands, xv–xvii. Washington, D.C.: Smithsonian Institution Press.

Nealon, Jeffrey T. 2008. *Foucault Beyond Foucault: Power and Its Intensifications Since 1984*. Stanford, Calif.: Stanford University Press.

Norton, Bryan G., Michael Hutchins, Elizabeth F. Stevens, and Terry L. Maple, eds. 1995. *Ethics on the Ark: Zoos, Animal Welfare, and Wildlife Conservation*. Washington, D.C.: Smithsonian Institution Press.

Novak, Melinda A., Peggy O'Neill, Sue A. Beckley, and Stephen J. Suomi. 1994. "Naturalistic Environments for Captive Primates." In *Naturalistic Environments in Captivity for Animal Behavior Research*, edited by Edward F. Gibbons Jr., Everett J. Wyers, Everett Waters, and Emil W. Menzel Jr., 236–258. Albany: State University of New York Press.

Olney, P. J. S., G. M. Mace, and A. T. C. Feistner, eds. 1994. *Creative Conservation: Interactive Management of Wild and Captive Animals*. Dordrecht: Springer.

Pádua, José Augusto. 2013. "Nature and Territory in the Making of Brazil." Translated by Rocky Hirst. In "New Environmental Histories of Latin America and the Caribbean," edited by Claudia Leal, José Augusto Pádua, and John Soluri. *RCC Perspectives: Transformations in Environment and Society*, no. 7:33–39.

Pissinatti, Alcides, Richard J. Montali, and Faiçal Simon. 2002. "Diseases of Lion Tamarins." In *Lion Tamarins: Biology and Conservation*, edited by Devra G. Kleiman and Anthony B. Rylands, 255–268. Washington, D.C.: Smithsonian Institution Press.

Reinert, Hugo. 2013. "The Care of Migrants: Telemetry and the Fragile Wild." *Environmental Humanities* 3:1–24.

Rose, Deborah Bird. 2004. *Reports from a Wild Country: Ethics for Decolonisation.* Sydney: University of New South Wales Press.

———. 2006. "What If the Angel of History Were a Dog?" *Cultural Studies Review* 12, no. 1:67–78.

Rylands, Anthony B., Jeremy J. C. Mallinson, Devra G. Kleiman, Adelmar F. Coimbra-Filho, Russell A. Mittermeier, Ibsen de Gusmão Câmara, Cláudio B. Valladares-Padua, and Maria Iolita Bampi. 2002. "A History of Lion Tamarin Research and Conservation." In *Lion Tamarins: Biology and Conservation,* edited by Devra G. Kleiman and Anthony B. Rylands, 3–41. Washington, D.C.: Smithsonian Institution Press.

Shepherdson, David J. 1994. "The Role of Environmental Enrichment in the Captive Breeding and Reintroduction of Endangered Species." In *Creative Conservation: Interactive Management of Wild and Captive Animals,* edited by P. J. S. Olney, G. M. Mace, and A. T. C. Feistner, 167–177. Dordrecht: Springer.

Shepherdson, David J., Jill D. Mellen, and Michael Hutchins, eds. 1998. *Second Nature: Environmental Enrichment for Captive Animals.* Washington, D.C.: Smithsonian Institution Press.

Snowdon, Charles T. 1994. "The Significance of Naturalistic Environments for Primate Behavioral Research." In *Naturalistic Environments in Captivity for Animal Behavior Research,* edited by Edward F. Gibbons Jr., Everett J. Wyers, Everett Waters, and Emil W. Menzel Jr., 217–235. Albany: State University of New York Press.

Snyder, Noel F. R., Scott R. Derrickson, Steven R. Beissinger, James W. Wiley, Thomas B. Smith, William D. Toone, and Brian Miller. 1996. "Limitations of Captive Breeding in Endangered Species Recovery." *Conservation Biology* 10, no. 2:338–348.

Spotte, Stephen. 2006. *Zoos in Postmodernism: Signs and Simulation.* Madison, N.J.: Fairleigh Dickinson University Press.

Stengers, Isabelle. 2000. *The Invention of Modern Science*. Translated by
Daniel W. Smith. Minneapolis: University of Minnesota Press.

Stoinski, Tara S., Benjamin B. Beck, M. A. Bloomsmith, and Terry
L. Maple. 2003. "A Behavioral Comparison of Captive-Born, Reintroduced Golden Lion Tamarins and Their Wild-Born Offspring." *Behaviour* 140, no. 2:137–160.

Stoinski, Tara S., Benjamin B. Beck, M. Bowman, and J. Lenhardt. 1997.
"The Gateway Zoo Program: A Recent Initiative in Golden Lion
Tamarin Reintroductions." In *Primate Conservation: The Role of Zoological Parks*, edited by Janette Wallis, 113–129. Norman, Okla.: American Society of Primatologists.

Strum, Shirley C., and Linda M. Fedigan, eds. 2000. *Primate Encounters:
Models of Science, Gender, and Society*. Chicago: University of Chicago
Press.

Tudge, Colin. 1992. *Last Animals at the Zoo: How Mass Extinction Can Be
Stopped*. Oxford: Oxford University Press.

Turner, Stephanie S. 2007. "Open-Ended Stories: Extinction Narratives in Genome Time." *Literature and Medicine* 26, no. 1:55–82.

van Dooren, Thom. 2016. "Authentic Crows: Identity, Captivity and
Emergent Forms of Life." *Theory, Culture and Society* 33, no. 2:29–52.

van Dooren, Thom, and Deborah Bird Rose. 2012. "Storied-Places in a
Multispecies City." *Humanimalia: A Journal of Human/Animal Interface
Studies* 3, no. 2:1–27.

Yamamoto, Maria Emília, and Anuska Irene Alencar. 2000. "Some
Characteristics of Scientific Literature in Brazilian Primatology." In
Primate Encounters: Models of Science, Gender, and Society, edited by Shirley
C. Strum and Linda M. Fedigan, 184–193. Chicago: University of
Chicago Press.

"Elliot's Bird of Paradise (*Epimachus ellioti*)." (Lithograph from Richard Bowdler Sharpe, *Monograph of the Paradiseidæ, or Birds of Paradise, and Ptilonorhynchidae, or Bower-Birds* [London: Sotheran, 1891–1898], 1:plate XVI. Reproduced by the National Library of Australia)

3. EXTINCTION IN A DISTANT LAND

The Question of Elliot's Bird of Paradise

RICK DE VOS

NATURE'S CHOICEST TREASURES

In 1824, the French naturalist René Primevère Lesson, in his capacity as pharmacist and botanist on Louis Duperrey's circumnavigatory voyage aboard *La Coquille*, arrived in western New Guinea in search of wildlife. There he encountered a bird of paradise for the first time:

> The view of the first Bird of Paradise was overwhelming. The gun remained idle in my hand for I was too astonished to shoot. It was in the virgin forest surrounding the harbour of Dorey. As I slipped carefully along the wild pigs' trails through this dusky thicket, a *Paradisea* suddenly flew in graceful curves over my head. It was like a meteor whose body, cutting through the air, leaves a long trail of light. With the ornamental plumes pressing against its flanks, this bird resembles an ornament dropped from the curls of a gouri and floating idly in the layer of air that encircles our planet's crust. (quoted in Gilliard 1969:23)

Having recovered from his initial astonishment, Lesson spent two weeks observing birds of paradise, shooting and collecting specimens of the Lesser and the King species.

Thirty years later, Alfred Russel Wallace began his travels through New Guinea and the Papuan islands, collecting natural history specimens for his private collection and for sale to museums and other collectors. Between 1854 and 1862, he encountered villages and smaller settlements, as well as vast mountain and jungle areas. In *The Malay Archipelago* (1869), he describes in great detail the hardships he endured in New Guinea and the surrounding islands, emphasizing the effort needed to make headway through the difficult terrain. He believed himself at the time to be the only Englishman to have seen birds of paradise in their native jungles and forests and to have obtained specimens from such environments. Wallace was acquainted with Lesson's work, and he observed and collected specimens of the Greater, Lesser, Red, and King Birds of Paradise, as well as describing and naming Wallace's Standard Wing. Wallace ([1869] 1962) expressed disappointment at not having identified more specimens, having been informed that rarer species could be obtained only several days' journey into the interior of the island:

> It seems as if Nature had taken precautions that these her choicest treasures should not be made too common, and thus be undervalued. . . . The country is all rocky and mountainous, covered everywhere with dense forest, offering in its swamps and precipices and serrated ridges an almost impassable barrier to the unknown interior; and the people are dangerous savages, in the lowest stage of barbarism. In such a country, and among such a people, are found these wonderful productions of Nature, the Birds of Paradise, whose exquisite beauty of form and colour, and strange developments of plumage are calcu-

lated to excite the wonder and admiration of the most civilized and the most intellectual of mankind, and to furnish inexhaustible materials for study to the naturalist, and for speculation to the philosopher. (439)

In 1873, while picking his way through a shipment of bird skins and plumes that he had imported from Singapore, the London taxidermist Edwin Ward found the plume of a male bird that he was unable to identify. His attention was caught by the depth of its coloring, noting that the specimen's back and tail were "beautifully illuminated with an amethyst colour" (Fuller 1995:56). He exhibited it before the Zoological Society of London in the same year, naming it Elliot's Bird of Paradise (*Epimachus ellioti*), after the American author and artist Daniel Giraud Elliot. The ornithologist and artist John Gould bought the specimen, which was sent to the British Museum after his death in 1881 and which constitutes the type specimen.

The significance of Ward's discovery appeared to be confirmed a few years later, in 1890, when another specimen identified as Elliot's Bird of Paradise was received by Adolf Bernard Meyer, the director of the Staatliches Museum für Tierkunde in Dresden. Meyer commented on the iridescent quality of the specimen's back and tail feathers, noting that the female was unknown and suggesting that the specimen came from northwestern New Guinea, possibly the island of Waigeo.[1] In 1930, the German ornithologist Erwin Stresemann inspected both specimens and declared Elliot's Bird of Paradise to be a hybrid rather than a "real" species. Ornithologists have continued to express doubts about the bird's taxonomic status, however. As recently as 2012, Julian Hume and Michael Walters have suggested that it is likely that the elusive bird is "either rare or extinct, occurring in a restricted montane range in New Guinea" (342).

These historical vignettes encapsulate many of this chapter's central concerns. Lesson's and Wallace's romantic accounts of their encounters and Ward's and Meyer's vivid specimen descriptions typified the way that the island of New Guinea and birds of paradise were represented to a European public: as ethereal, exotic, beautiful, and seductive. The color and emotion conveyed in their descriptions of the distant and unfamiliar resonated with an industrialized Europe in search of an escape from its increasingly ordered spaces and times. Stories of inaccessible terrain and fierce natives ran parallel with those of the rewards that were there for the taking for those who dared to venture to the islands: the promise of sublime beauty, commercial and political power, and scientific discovery.

The fate of Elliot's Bird of Paradise exemplifies the way in which the lives and deaths of rare birds of paradise have been obscured by persistent colonial representations of New Guinea and its wildlife. The narrative of a paradise beckoning hunters and collectors holds no place for stories of dispossession and extinction. Just as historical and geographical representations of New Guinea continue to depict the island as a space of colonial desire, scientific representations of birds of paradise persist in conflating a subjective, aesthetic response to the plumage and behavior of the birds with an implied scientific objectivity. In the late nineteenth and twentieth centuries, European representations of New Guinea started to emphasize it as a space of production, a space in which the slaughter of birds of paradise for the plume trade was justified and written over. This chapter argues that the cultural demands of history, commerce, and science have worked to conceal and deny the possible loss of bird of paradise species, and the conditions leading to that loss.

West Papua and Papua New Guinea

The island of New Guinea and the small islands surrounding it are divided politically into two distinct regions. The western half of the island and the small island groups surrounding it constitute the Indonesian provinces of Papua and West Papua, known collectively as West Papua, and the eastern half of the island constitutes the nation of Papua New Guinea. The majority of the human population in both West Papua and Papua New Guinea are ethnic Papuan, with people of Austronesian origin, European expatriates, and more recent Indonesian migrants making up most of the rest of the population. However, such categorization belies the complexity of society and culture in New Guinea. With over 1,000 different tribal groups and languages spread across the island, New Guinea is the most ethnically and linguistically diverse geographical area in the world, challenging any notion of a homogeneous, knowable entity (Foley 1986:3).

New Guinea's terrestrial ecological regions range from montane rain forest to savanna and grasslands to mangroves, displaying a vast level of biodiversity with a high percentage of endemic plants and animals. Many parts of New Guinea have been neither seen nor visited by outsiders, and it is estimated that there are many thousands of species of insects and plants and hundreds of species of birds that have not been identified or recorded by scientists or other visitors (Beehler 2007:9–10).

From the sixteenth century, European countries have perpetrated many violent military and colonial incursions in New Guinea. Portuguese sailors, having traded bird of paradise plumes with Malay merchants, sighted and laid claim to parts of New Guinea between 1511 and 1529. Spanish trading posts were established as part of the Spanish East Indies between 1545 and 1606. In 1660, the Dutch East India Company claimed the island from the Spanish, who had been unable to maintain settlement there and formally relinquished claims to New Guinea in 1715. In 1793, the entire island was claimed for Great Britain by the East India Company. The claim was disputed by the Netherlands, and in 1828 the Dutch East India Company took possession of the western half of New Guinea (Gilliard 1969:18–19). Since that time, parts of the island have been subject to claims, settlement, and military action by Germany, Great Britain, the Netherlands, Japan, Australia, and Indonesia.

West Papua and Papua New Guinea have separate recent histories but are connected ecologically and by sustained experiences of colonial and military violence, as well as by global commercial pressures.[2] While governance and environmental-management problems are different in Papua New Guinea and West Papua, the need to address environmental degradation, habitat loss, and the overexploitation of resources is a critical one for the whole of the island. Environmental pressures and threats include land degradation due to intensive agriculture and livestock grazing, unsustainable logging practices, large-scale mining operations, and increased population pressure.

A LAND RICH AND STRANGE

While Papuan people have lived on the island of New Guinea for at least 40,000 years, practicing agriculture and trading with Malay and Indonesian seafarers since at least the twelfth century, European representations of New Guinea have perceived the island as remote, exotic, strange, and unknown. The inhabitants of New Guinea were portrayed as primitive headhunters, lost in time and space. Since the time the Spanish explorer Yñigo Ortiz de Retes arrived in the north of the island in 1545, claiming the island for Spain and naming it Nuevo Guinea, New Guinea came to be defined in relation to other people and other spaces (Trotter

1884:197). As was the case in so many sites of colonization, an un-known space was named and given meaning in relation to a known one, emptied of its intrinsic meanings, and painted over to reflect the vision of the colonizers. For Ortiz de Retes, the Papuan peo-ple resembled the inhabitants of the Spanish colonies in West Africa known collectively as Guinea. Thus they represented a new version of colonial subjects he knew little about. New Guinea constituted a conquest and a point in a larger itinerary of explora-tion and possession. Other Europeans projected their own cultural and commercial desires onto what they envisioned as a smaller version of their image of Africa, replete with jungles and savage inhabitants. The name New Guinea marked a wholly other-defined space, a reflection of a reflection that itself remained obscure. The island subsequently became a calling place for many explorers. Their reports, as well as a growing scientific interest in the re-gion, led to exploration by a number of private and government expeditions. However, the interior of New Guinea remained largely free from European explorers, scientist, and traders until the middle of the twentieth century (Gilliard 1969:22). A common experience of European visitors was of finding the mountain for-ests inaccessible, particularly in the western and northern parts of the island (Frodin and Gressitt 1982:89). This lack of access—combined with the strangeness of the terrain, the people, and the wildlife—led to much conjecture as to what lay beyond the coastal areas. New Guinea thus came to be imagined through previous colonial conquests and the lure of the unknown.

DISPLAY AND SEDUCTION

Birds of paradise and people have coexisted for most of New Guinea's long history of human habitation. Papuan people have studied the courting and feeding behavior of the birds, taking note of the fruiting trees visited by the birds and adopting the movements of certain species in traditional dances and rituals. Within specific language groups, detailed systems of taxonomy based on appearance, song, habitat, and behavior were employed to describe and understand this diverse group of birds (Healey 1993:21). Prior to European colonial activities in the nineteenth century, Papuan people observed tighter territorial boundaries, with strict protocols regarding moving from one cultural area to another. Consequently, birds of paradise with limited ranges might come into contact with only a small group of people (Frith and Frith 2010:87). Adult male birds were hunted and killed, their skins and plumes used for ceremonial purposes and personal display, and as a part of bride-prices and exchanges between clans. Papuan people equated the colorful plumage and elaborate courting dances of some male birds of paradise with a strong sense of virility and attractiveness to women. For example, the Maring-speaking people from Papua New Guinea's central highlands identify culturally with the Black Sicklebill Bird of Paradise (*karanc*), the nuptial male plumage of which is glossy black rather than brightly colored. *Karanc* are also noted, however, for their iridescent flank plumage, prominent curved beaks, and long and spectacular sets of curved tail feathers. The dark, shimmering plumes of *karanc* are associated with the healthy skin of virile Maring males. Young Maring men wear *karanc* plumes and engage in collective courting dances based on the sustained, energetic dances of male *karanc* in order to attract young women (Healey 1993:29–30).

This long history of intimate interest in avian display, however, was redirected and incorporated into European colonial and capitalist ventures. Papuan people's connections with and knowledge of birds of paradise were exploited by plume traders, who employed local hunters to kill, skin, and preserve the birds, trading the skins for axes, steel knives, and tobacco (Kirsch 2006:19). Traders sold the skins to wealthy private collectors, capitalizing on the growing demand for skins and mounted specimens, particularly those deemed rare or possessing striking plumage, for display in cabinets. Collectors were afforded a degree of prestige and social status as a result of their collections. The demand for specimens for museums, as well as live birds for public and private aviaries, also intensified between the early and mid-nineteenth century. Despite this, very few collectors or ornithologists had ever seen live birds of paradise, and taxidermists and artists used a great deal of license in imagining them from plumes and skins (Frith and Frith 2010:131). The late-eighteenth- and nineteenth-century vogue for collecting and displaying natural history specimens nonetheless provided the basis of European scientific knowledge about these birds.

The paucity of firsthand knowledge of birds of paradise in Europe led to increased speculation about them. Oil and watercolor paintings outdid one another in envisioning the allure of the males' plumes. Part of the cultural work of these paintings was to write over the many local relationships the birds were a part of and the knowledge acquired over time within these intraspecies and interspecies relationships. The paintings depicted birds of paradise as discrete individuals or pairs set against white or pale backgrounds, removed from any reference to the natural environments they inhabited, with the male birds foregrounded as colorful, collectable, and fashionable. Daniel Elliot's *A Monograph of the Paradiseidæ, or Birds of Paradise* (1873), with illustrations by Joseph Wolf; John Gould's *The Birds of New Guinea and the Adjacent Papuan Islands* (1875–1888), illustrated by Gould; and Richard Bowdler Sharpe's *Monograph of the Paradiseidæ, or Birds of Paradise, and Ptilonorhynchidae, or Bower-Birds* (1891–1898), with illustrations by Sharpe, Gould, William Hart, and John Gerrard Keulemans all presented spectacular paintings of birds of paradise that captured the imagination of late-nineteenth-century Europeans. The illustrations became the centerpieces and principal selling points of the volumes. The paintings reflected the desires for the exotic and new, as well as for order and classification. These influential texts nurtured the growing European taste for bird paintings and sculpture, and for textiles and fashion items inspired by the extraordinary shapes and vibrant colors of bird of paradise plumes (Frith and Frith 2010:138).

Paul Farber (1980:392) has argued that as European scientific interest in birds increased in the nineteenth century, and as European museums sought to increase their collections of specimens, the collectors' overriding concern with taxonomy and classification endured, continuing to shape ornithology as it developed into a more specialized science. Colonization and exploration maintained and reinforced ornithology's inordinate attention to the exotic and the unknown. Bird of paradise specimens, with males resplendent in their nuptial displays, brought color, distinctiveness, and drama to collectors' cabinets. Their obsession with male birds and their plumage influenced the way information about particular species was collected and documented. Polygynous birds of paradise show an extreme degree of sexual dimorphism, with adult males developing a wide range of display features, especially elaborate, colorful and elongated head, wing, and tail feathers. Female birds of paradise tend to be far more homogeneous and less conspicuous in appearance (Attenborough and Fuller 2012:12). Plume hunters paid little attention to female and juvenile birds, and collectors had negligible access to them. Consequently, European naturalists and collectors experienced difficulties in identifying and differentiating among female birds of paradise.

Taxonomic practices in the nineteenth and early twentieth centuries focused primarily on morphology. Ornithologists and naturalists concentrated on the appearance of the birds, particularly adult males' nuptial plumage, with most unable to study birds of paradise in situ. Observations and anecdotes from field collectors regarding the birds' elaborate courting behavior supplemented these visual studies. Charles Darwin ([1859] 1968) discussed birds of paradise as furnishing extreme examples of sexual selection, a conclusion related to the diversity of their structure, color, and ornamentation:

> [B]irds of paradise . . . congregate; and successive males display their gorgeous plumage and perform strange antics before the females, which standing by as spectators, at last choose the most attractive partner. . . . [I]f a man can in a short time give elegant carriage and beauty to his bantams, according to his standard of beauty, I can see no good reason to doubt that female birds, by selecting, during thousands of generations, the most melodious or beautiful males, according to their standard of beauty, might produce a marked effect. I strongly suspect that some well-known laws with respect to the plumage of male and female birds, in comparison with the plumage of the young, can be explained on the view of plumage having been chiefly modified by sexual selection. (137)

Fatefully, it was these very features of birds of paradise, their elaborate appearance and behavioral displays eliciting desire and seduction, which Darwin saw as exemplary of evolutionary development and sexual selection, that were ultimately to evince an imperial desire for the birds' alluring plumes and to furnish the means by which they were violently extracted.

THE "PLUME BOOM"

The sexual behavior of birds of paradise, particularly court-arena displays, brought with it connotations of attraction and desire for European society. While Papuan people culturally associated birds of paradise—especially adult males in full plumage—with youth and masculinity, Europeans increasingly associated the birds and their plumage with the feminine, taking the striking beauty and exotic allure of the plumage as seen in cabinet collections, books, and paintings and projecting them onto the bodies of affluent European women. As British, German, and Dutch milliners and fashion designers, their imagination and enterprise fueled by their countries' colonial exploits, turned their attention toward bird plumes in the late nineteenth century, the demand for birds from New Guinea increased exponentially (Frith and Frith 2010:108). London, Berlin, and Amsterdam became the centers of a burgeoning millinery industry focused on bird plumes, with bird of paradise plumes in particular demand due to their beauty, rarity, and expense. Milliners produced hats decorated with the feathers, wings, and entire bodies of birds of paradise. The routes that had been established for the trade of skins for scientific study soon became the routes for the trade of plumes for millinery and fashion, with illustrated catalogs allowing buyers in Europe to order specific plumes for hats (Kirsch 2006:16–17). The desire evoked in bird of paradise paintings was transferred to a more explicit site of economic acquisition and consumption, with catalogs displaying single plumes and feathers amplifying the reductive representation of ornithological illustrations and removing any sense of the birds as living beings connected to other forms of life.

While no precise figures exist detailing the slaughter of birds of paradise during this period, it is estimated that many thousands of birds were shot each year between 1875 and 1914. Stuart Kirsch (2006:16) estimates that between 1905 and 1920, 30,000 to 80,000 bird of paradise skins were exported annually to the feather auctions of London, Paris, and Amsterdam. The nuptial displays that initiated the cycle of new life for birds of paradise now contributed to their demise. Foreign hunters, dependent on the knowledge of local Papuans in maximizing the number of adult male birds they could find to kill, were led to areas where mature males participated in communal courtship displays and were thus distracted and vulnerable to shooting.

Whereas Papuan hunters shot birds of paradise with arrows, shotguns were used by European, Malay, Chinese, and Australian hunters to "hunt plumes" and "collect specimens." Papuan hunters gradually replaced their arrows with shotguns. Clifford Frith and Bruce Beehler (1998) contend that the use of shotguns enabled Papuan hunters other than traditional landowners to kill and take birds from prohibited land without detection and punishment. It also removed the need to gradually learn the natural history of particular species and their patterns of behavior. Shotguns changed the time and space involved in hunting the birds. Traditional hunters, dependent on bows and arrows, needed to acquire a detailed knowledge of the lives and relationships of the birds. Arrows were precious, and a wasted shot meant that no birds could be hunted in the immediate future. The shotgun enabled birds to be killed in short periods of time, in rapid succession, without the need to build up this knowledge and gain a sense of living with the birds. While individual male birds were targeted by hunters, a shotgun blast could incidentally kill other birds in close proximity, including females and juveniles. In some cases, shooting the birds also meant controlling the activities of local Papuan hunters, utilizing their knowledge of their surroundings and of the birds that shared their space. The relationship between birds of paradise and Papuan people was irrevocably changed.

The indiscriminate nature of the killing and the removal of the skins and plumes to European warehouses without documentation meant that the origins of the birds killed could not be traced. While European ornithologists drew many benefits from the plume trade in terms of the availability of skins and plumes, Richard Bowdler Sharpe in 1891 lamented the lack of biological data collected on the nature and location of wild bird of paradise populations: "[T]hat we shall ever discover them can scarcely be expected, for the aim of every ordinary collector in the present day seems to be, not to furnish us with details of the nesting habits of the Birds of Paradise, but to see how many of these beautiful creatures he can procure for the decoration of the hats of the women of Europe and America" (quoted in Frith and Beehler 1998:42).

E. Thomas Gilliard (1969:23) contends that while the visits of naturalists to New Guinea have been well documented since the middle of the nineteenth century, the discoveries of thousands of plume collectors have not. Professional ornithologists meticulously picked over shipments of bird of paradise plumes bound for commercial trade in Europe, saving rare specimens for museums and private collections, but the geographical origins of the plumes were obscure or unknown. Historical documentation of the unprecedented killing of birds of paradise and the plumage trade is extremely limited. Foreign hunters were not keen to publicize the locations, methods of access, and details of their acquisition of plumes, and, unlike some zoological museum specimens and taxidermy mounts, millinery skins and plumes were rarely well preserved and were generally disposed of once the fashion had passed (Patchett 2012:19).

Fatefully, hunters started noticing the rapid depletion in numbers of sought-after bird of paradise species. Greater Birds of Paradise and Black Sicklebills were in particular demand, and adult males became increasingly difficult to find (Gilliard 1969:134). The cumulative killing of millions of birds of paradise in the late nineteenth and early twentieth centuries brought about the disappearance of birds from areas previously visited for court display and, with it, speculation regarding the survival of certain species (Patchett 2011; Swadling 1996:252–255). In the 1920s, all birds of paradise species were protected from export out of New Guinea, but by that time European millinery fashions had begun to turn away from hats with plumes. In Great Britain, the Plumage League was formed in 1889, in order to campaign against the worldwide killing of birds for the fashion industry, eventually receiving a royal charter and becoming in 1904 the Royal Society for the Protection of Birds (Kirsch 2006:19). Similar organizations were established in other European countries, and their efforts were instrumental in legal bans and restrictions on the killing of birds and the importing of plumes. With birds of paradise faced with the threat of extinction, campaigning against the killing of birds for plumes appeared to become more fashionable than wearing their plumes, and perhaps it is not surprising that efforts by public and community groups in Europe to stop plume hunting and the plume trade have received far greater documentation than the killing and trade itself. Nevertheless, this colonial extraction, driven by a desire for the exotic, had devastating effects on bird of paradise numbers, resulting in the near extinction of a number of species.

BIRDS AND FRUIT

The relationship between species of polygynous birds of paradise—that is, species in which males seek to mate with more than one female partner—and particular fruiting trees in New Guinea provides some clues as to the fate of both in a changing environment. Most bird of paradise species are predominantly frugivorous, many supplementing their diet with insects found on tree bark and branches. Numerous studies of polygynous birds have addressed the hypothesis that frugivory provides birds with an accessible and nutrient-rich diet, enabling them to spend more time engaged in behaviors related to sexual selection (for example, Gilliard 1969; Snow 1980). Beehler (1983) postulates that frugivory also promotes larger, overlapping home ranges with the potential for greater intraspecies and interspecies contact between individual birds, suggesting that "aspects of this coevolutionary relationship between birds of paradise and some fruiting plants may contain an answer to why the birds were able to evolve their remarkable polygamous mating systems" (9).

The canopies of fruit trees provide sites for birds of paradise to prepare and maintain court arenas, where males engage in vocal and physical performances in order to attract females. Jared Diamond (1986:23) argues that polygynous birds of paradise tend to feed on a wide variety of fruits, particularly capsulate or drupaceous fruit. Trees with these fruits have a much smaller range of avian visitors, with some visited exclusively by birds of paradise. Many capsular fruits have to be pulled apart and pecked open with a strong beak in order to yield the fleshy, brightly colored seeds held within, a task at which birds of paradise are particularly adept. Through these activities, birds of paradise play an important role in seed dispersal and germination for these trees.

In Beehler's (1983) study, *Chisocheton weinlandii*, a fruiting tree with capsular fruit, was visited frequently and exclusively by birds of paradise in a small steady stream, with each bird rarely taking more than two or three fruit per visit. These fruits are richer in proteins and lipids than are figs. Henry Howe (1984:274) has suggested that particular species of birds of paradise may prove pivotal for the survival of particular tree species. Indeed, several species of *Chisocheton* endemic to New Guinea have been classified as either vulnerable or endangered by the International Union for Conservation of Nature (2015). While habitat loss through environmental degradation is an important contributing factor, a decline in bird of paradise numbers may well have played a part in reducing the range and distribution of tree species, and the extinction of a tree species may have been caused by the extinction of a mutualist bird species.

HYBRIDITY

An identified characteristic of a large number of polygynous bird of paradise species is a propensity to interspecific and intergeneric hybridization in the wild, suggesting that these species are likely to be closely related despite being classified in different genera. Frith and Beehler (1998:501) speculate that hybridization as a phenomenon among polygynous court-displaying birds of paradise is now accepted as prevalent, not because of physical characteristics but for behavioral reasons: the young of such birds have little or no contact with adult males, and intrasexual competition among males focuses on advertisement, display, and copulation, meaning that males are likely to mate with any females that respond favorably to their displays.

Questions regarding whether particular birds of paradise constitute species or hybrids stem from their relative rarity. Hybrid birds of paradise have been identified in Europe mainly through the arrival of commercial-trade skins and collectors' specimens. Most of them were originally considered to be new species, until a suggestion by the German ornithologist Anton Reichenow in 1901 that a bird he had originally described as a new species displayed characteristics that indicated a possible cross between two other species of bird of paradise in the wild (Frith and Beehler 1998:499). The ornithological obsession with taxonomy led to much debate regarding the classification of particular specimens. In 1930, the highly respected German ornithologist Erwin Stresemann conducted a review of all the bird of paradise species that had not been seen in the wild and contended that all of them were hybrids. In a paper published in the journal *Novitates Zoologicae*, he identified putative parent species for each bird known from only museum specimens. Writing in 1954, he stated: "I decided in 1930 to examine all 'suspicious' species of Birds of Paradise to see whether they might prove to be generic or specific hybrids. I reached the surprising conclusion that no less than 18 species and 8 genera should be removed from the list of 'normal' Birds of Paradise because they were hybrids. Time has proved that I was right" (quoted in Gilliard 1969:63).

While his findings were widely accepted, some ornithologists expressed doubts as to whether every single specimen could be incorporated into Stresemann's schema, suggesting that some specimens were just as likely to represent rare or extinct species.[3] In *The Lost Birds of Paradise* (1995), Errol Fuller examines the history of each of the bird of paradise specimens that Stresemann identified as constituting hybrids rather than species. While Stresemann paid close attention to physical characteristics and, to some extent, geographical distribution, he did not take behavior, particularly court-display details, into account. Fuller contends that while Stresemann was probably correct in some of his diagnoses, others were speculated on without adequate substantiation. Half a dozen, according to Fuller (1995:18), are just as likely on existing evidence to constitute legitimate species that are now lost or extinct. For example, as previously noted, Stresemann proposed that Elliot's Bird of Paradise is a hybrid, naming the Black Sicklebill Bird of Paradise and the Arfak Astrapia as putative parent species. However, Fuller argues that the Astrapia is a fanciful choice, with little supporting evidence, and that Elliot's Bird of Paradise is much smaller than the two proposed parent species. The specimens also show a number of characteristics not present in either parent species, adding weight to the possibility of the specimens constituting a unique species (Fuller 1995:64).[4]

Stresemann's declaration of specimens as representative of hybrids rather than species on the basis of their appearance draws attention to the function of death and absence in classifying living things. The enunciation of hybridity within the discourse of biological science constitutes a writing practice undertaken in a space and time "after" the life and death of the subject being observed and classified.[5] In the case of birds of paradise deemed to be hybrids, the subjects are dismissed as a presence without significance. The work of Stresemann exemplifies how, in taxonomic practice, hybridity is recorded in the absence of the specimen's spatial, temporal, and familial context, an act of disavowal. For ornithologists in the late nineteenth and early twentieth centuries— focused on anatomy, appearance, and classification rather than on the observation of live birds—inscribing hybridity, rarity, or extinction was dependent on birds being killed and delivered for study. As a response to a gap in knowledge, hybridity ultimately works to deny both speciation and extinction, and thus the anthropogenic activities contributing to the latter. The conceptualization and enunciation of species and hybridity work together so as to determine that the lives of some birds are more significant and more valuable than the lives of others and, by extension, the absence of their lives and ways of living. In the case of birds of paradise in the late nineteenth and early twentieth centuries, this enunciation helps to exculpate the perpetrators of the widespread massacre, downplaying the disappearance of particular birds as the loss of hybrids rather than the extinctions of species. The significance of hybridity in the case of Stresemann's study is tied to the nuptial plumes of adult male birds, rather than to whole species. Younger males and females may still exist and perpetuate the species in the absence of older males. However, an equally likely possibility is that a specific group of birds, with their particular experiences and way of life, has disappeared forever from the world.

Space, Time, and Extinction

It seems sad that on the one hand such exquisite creatures should live out their lives and exhibit their charms only in these wild inhospitable regions, doomed for ages yet to come to hopeless barbarism; while, on the other hand, should civilized man ever reach these distant lands, and bring moral, intellectual, and physical light into the recesses of these virgin forests, we may be sure that he will so disturb the nicely-balanced relations of organic and inorganic nature as to cause the disappearance, and finally the extinction, of these very beings whose wonderful structure and beauty he alone is fitted to appreciate and enjoy. (Wallace [1869] 1962:448–449)

Accompanying the violent takeover and exploitation of other territories between the fifteenth and twentieth centuries, European societies produced and imposed a notion of colonial space, in order to justify their particular political, economic, and social activities.[6] Colonial space utilizes art, literature and other cultural products in reinforcing colonial activities as productive and desirable. From the middle of the sixteenth century, New Guinea was imagined and realized by Europeans as a colonial space, one of political and economic potential. British, German, and Dutch attempts to claim and settle the island shaped the way New Guinea was represented globally and in everyday life in Europe. Overwriting centuries of trade with Malay and Southeast Asian sailors, and the violence involved in establishing colonial hierarchies, New Guinea came to be seen as an exploitable space connected to Europe, a frontier signifying adventure and potential reward. Birds of paradise were represented as ethereal and alluring inhabitants of this remote, exotic land—indeed, part of the treasure the island promised. Collectors' cabinets, oil paintings, and illustrated books worked together in depicting the birds as sublimely sensuous and beautiful.

By the nineteenth century, this representational space had again been transformed, this time in relation to a modern, industrialized Europe. New Guinea was recast as a space of production, now inextricably linked to its equivalents in Europe: millinery factories, fashion warehouses, and feather auctions. The work of European plume hunters and traders, fashion designers and manufacturers concealed the labor of Papuan people who possessed knowledge of the locations of bird of paradise courting grounds and who were employed to bring hunters to these areas. Beautiful, fashionable hats became a commodity that both justified and obscured their own production, invoking, rather, the beauty and rarity of the birds: "Space is liable to be eroticised and restored to ambiguity, to the common birthplace of needs and desires, by means of music, by means of differential systems and valorisations which overwhelm the strict localisations of needs and desires in spaces specialised either physiologically (sexually) or socially (places set aside, supposedly, for pleasure)" (Lefebvre 1991:391).

Representations of the plume and hat industry in the form of fashion catalogs, parades, and advertisements overlaid stories of Papuan savagery and the realities of the large-scale slaughter of birds. The bodies of birds of paradise became discursively and figuratively hollowed out and dismantled in this spatial practice, leaving skins and feathers, signs of allure and desire achieved through the deaths of others. Birds of paradise were returned as plumes: signs of a transferable beauty and rarity. The deaths of the birds and the violence involved in transforming living birds into skins and plumes were sublimated within the language and logic of fashion and science.

These various colonial representations transformed New Guinea into a space that was simultaneously both knowable and unknown. Such depictions attracted explorers, naturalists, and commercial traders, each armed with different impressions of what the island promised. Private collectors and plume hunters had their reasons for keeping their knowledge of birds of paradise to themselves. Representations of birds of paradise as rare and exotic, shrouded in mystery, allowed a lack of knowledge about their distribution and behavior to be normalized and their slaughter and transformation into fashionable hats to continue until a growing global concern for the future of the birds led to the trade being stopped. The enduring notion of New Guinea as colonial space justified this time of killing as productive and fulfilling of a desire, proceeding without adequate documentation or consideration of its impact on bird populations. Feathers, plumes, and specimens resisting classification also mark this as a time of possible extinction, however, the knowledge of which has been confounded and denied.

Whether or not the birds known as Elliot's Birds of Paradise constituted a rare species or a small group of hybrid birds, their existence is irrefutable. They were likely to have lived in montane forests in the northwest of New Guinea, to have closely resembled other sicklebills, and to have had a largely frugivorous diet, with groups of male birds engaging in court-arena display behavior to attract mates. The fact that both Ward's and Meyer's specimens were found in the nineteenth century among skins destined for the plume trade suggests that if they did become extinct, plume hunting may have been the principal cause. The unprecedented time of killing known as the "plume boom," in which the restrictions observed by Papuan hunters were forgotten in the name of industry and profit, may have led to both a number of extinctions and increased incidences of hybridization. It was a time when polygynous birds of paradise were likely to have become particularly vulnerable to extinction, bereft of time to adapt or develop survival strategies, and it is the artifacts of this period that provide the best chance of tracing any possible extinctions. The killing of adult male birds begs questions regarding females and juveniles. What Stresemann regarded as exemplary of a larger scheme of hybridization may have signified an indeterminate point of crisis in the lives of the birds we choose to forget or to remember as Elliot's Bird of Paradise.

1. Julian Hume and Michael Walters (2012) describe the coloring of the specimen identified by Meyer as "almost black with purple and violet iridescence, markedly so on crown in contrast to the black forehead; on cheeks the gloss is green; throat velvet blackish-brown merging into wash of olive-green on breast; broad band of dull grape-red crosses this area; false wings tipped with blue" (342).

2. Eben Kirksey's (2012) fieldwork in West Papua addresses some of the complexities of colonialism in the present day, providing an account of the diversity of ways in which West Papuan people have negotiated and resisted Indonesian colonial rule in order to make political and personal gains.

3. Tom Iredale (1950:5) dismisses Stresemann's review as premature, fanciful, and too readily accepted by the evolutionary biologist Ernst Mayr, who was Stresemann's pupil. While more measured in his response, E. Thomas Gilliard (1969) also expresses doubt that Stresemann was correct in every instance:

> [H]as time proved that Stresemann was right in all of his determinations? The final answer to this question I feel cannot be given until New Guinea has been fully explored. . . . Many of the hybrids involve high altitude species, and many high areas of New Guinea that, although visited in former times by Papuan plume collectors, have never been visited by ornithologists. Therefore, it is my belief that some of the birds which are now classified as hybrids are actually "lost" species that still await discovery. (63–64)

4. Errol Fuller (1995:32, 45) also raises serious doubts about Stresemann's identification of Rothschild's Lobe-Billed Bird of Paradise and Duivenbode's Rifle Bird as hybrids, arguing in each case that the putative parent species proposed by Stresemann were unlikely to come into contact with each other and that there was just as much evidence to

consider each a distinct species. Hume and Walters (2012:342, 344) support Fuller in each of these instances.

5. In describing hybridity as a form of writing, I am referring to Jacques Derrida's (1973:134, 1978:278–285) formulation of writing as all practices seeking to produce an inscription temporally and spatially removed from its object, such as visual art, music, photography, cinematography, modeling, genetic coding, and computer programming. Each of these practices constitutes a field of indeterminate traces and retentions rather than a definitive record of the object being inscribed. These writings are never equivalent to their objects: they may replace, add to, or subtract from their object, but they always constitute a spatial and temporal removal. Inscribing hybridity in the absence of living birds constitutes an act of writing, removed from the time and space of the birds, that retrospectively denies both speciation and extinction.

6. I am drawing here on the work of Henri Lefebvre (1991:287–289), who argues that a society's conceptions of space, both abstract and material, are socially produced, the result of an ongoing process that, although strategically concealed, is historically traceable. For Lefebvre, the space of knowledge, of epistemology, is the product of a political and institutional process—abstraction. "Abstract space" is reinforced by a society's representations of space and brought to bear, as a set of limitations, on representational spaces. Spaces of work not only reinforce the notion of productive activity but also situate such activity within a system and hierarchy of production. This dual function also has a bearing on the meaning of "space" as an abstract concept. Abstract space works toward rendering all spaces homogeneous and discrete, able to be compared and contrasted. It assumes forms that seemingly resolve contradictions between globally conceived space and "fragmented" space, the result of a multiplicity of procedures caused by the establishment of markets and the division of labor.

REFERENCES

Attenborough, David, and Errol Fuller. 2012. *Drawn from Paradise: The Natural History, Art and Discovery of the Birds of Paradise, with Rare Archival Art.* New York: HarperCollins.

Beehler, Bruce M. 1983. "Frugivory and Polygamy in Birds of Paradise." *Auk* 100, no. 1:1–12.

——. 2007. "Introduction to Papua." In *The Ecology of Papua, Part One*, edited by Andrew M. Marshall and Bruce M. Beehler, 9–10. Singapore: Periplus.

Darwin, Charles. (1859) 1968. *The Origin of Species.* Harmondsworth: Penguin.

Derrida, Jacques. 1973. *Speech and Phenomena: And Other Essays on Husserl's Theory of Signs.* Translated by David B. Allison. Evanston, Ill.: Northwestern University Press.

——. 1978. *Writing and Difference.* Translated by Alan Bass. London: Routledge and Kegan Paul.

Diamond, Jared. 1986. "Biology of Birds of Paradise and Bowerbirds." *Annual Review of Ecology and Systematics* 17:17–37.

Elliot, Daniel Giraud. 1873. *A Monograph of the Paradiseidæ, or Birds of Paradise.* London: Printed for the Subscribers, by the Author.

Farber, Paul Lawrence. 1980. "The Development of Ornithological Collections in the Late Eighteenth and Early Nineteenth Centuries and Their Relationship to the Emergence of Ornithology as a Scientific Discipline." *Journal of the Society for the Bibliography of Natural History* 9:391–394. doi:10.3366/jsbnh.1980.9.4.391.

Foley, William A. 1986. *The Papuan Languages of New Guinea.* Cambridge: Cambridge University Press.

Frith, Clifford B., and Bruce M. Beehler. 1998. *The Birds of Paradise.* Oxford: Oxford University Press.

Frith, Clifford B., and Dawn W. Frith. 2010. *Birds of Paradise: Nature, Art, History.* Malanda, Australia: Frith & Frith.

Frodin, D. G., and J. L. Gressitt. 1982. "Biological Exploration of New Guinea." In *Biogeography and Ecology of New Guinea*, edited by J. L. Gressitt, 87–130. The Hague: Junk.

Fuller, Errol. 1995. *The Lost Birds of Paradise*. Shrewsbury: Swan Hill Press.

Gilliard, E. Thomas. 1969. *Birds of Paradise and Bower Birds*. London: Weidenfield & Nicolson.

Gould, John. 1875–1888. *The birds of New Guinea and the adjacent Papuan Islands, including many new species recently discovered in Australia* [Completed after the author's death by R. Bowdler Sharpe]. London: Sotheran.

Healey, Christopher. 1993. "Folk Taxonomy and Mythology of Birds of Paradise in the New Guinea Highlands." *Ethnology* 32, no. 1:19–34.

Howe, Henry F. 1984. "Implications of Seed Dispersal by Animals for Tropical Reserve Management." *Biological Conservation* 30:261–281.

Hume, Julian P., and Michael Walters. 2012. *Extinct Birds*. London: Poyser.

International Union for Conservation of Nature. 2015. "*Chisocheton stellatus*." The IUCN Red List of Threatened Species. http://www.iucnredlist.org/details/38170/0.

Iredale, Tom. 1950. *Birds of Paradise and Bower Birds*. Melbourne: Georgian House.

Kirksey, Eben. 2012. *Freedom in Entangled Worlds: West Papua and the Architecture of Global Power*. Durham, N.C.: Duke University Press.

Kirsch, Stuart. 2006. "History and the Birds of Paradise: Surprising Connections from New Guinea." *Expedition* 48, no. 1:15–21.

Lefebvre, Henri. 1991. *The Production of Space*. Translated by Donald Nicholson-Smith. Oxford: Blackwell.

Patchett, Merle. 2011. "Fashioning Feathers: Dead Birds, Millinery Crafts and the Plumage Trade" [museum exhibition]. https://fashioningfeathers.info/about/.

——. 2012. "On Necro-Ornithologies." *Antennae* 20:9–26.

Sharpe, Richard Bowdler. 1891–1898. *Monograph of the Paradiseidæ, or Birds of Paradise, and Ptilonorhynchidae, or Bower-Birds*. London: Sotheran.

Snow, D. W. 1980. "Regional Differences Between Tropical Floras and the Evolution of Frugivory." In *Acta XVII Congressus Internationalis Ornithologici*, edited by R. Nohring, 1192–1198. Berlin: Deutschen Ornithologen-Gesellschaft.

Swadling, Pamela. 1996. *Plumes from Paradise: Trade Cycles in Outer Southeast Asia and Their Impact on New Guinea and Nearby Islands Until 1920*. Boroko: Papua New Guinea National Museum.

Trotter, Coutts. 1884. "New Guinea: A Summary of Our Present Knowledge with Regard to the Island." *Proceedings of the Royal Geographical Society and Monthly Record of Geography* 6, no. 4:196–216.

Wallace, Alfred Russel. (1869) 1962. *The Malay Archipelago: The Land of the Orang-Utan and the Bird of Paradise*. New York: Dover.

Patrick Ching, *Hawaiian Monk Seal* (*Monachus schauinslandi*), 1987.
(© Patrick Ching)

4. MONK SEALS AT THE EDGE

Blessings in a Time of Peril

DEBORAH BIRD ROSE

I was buzzing with enthusiasm when I arrived in Honolulu just before Christmas in 2011. The paperwork was finally in place, and now I had the opportunity to learn all I could about the critically endangered Hawaiian monk seals and the people who were dedicated to trying to ensure their future. The first port of call was the Waikiki Aquarium. Part of my excitement arose from the timing of this trip. The famous young monk seal KP2 had just returned to Hawai'i after a stint at the University of California, Santa Cruz, where he had been a research participant in the Marine Mammal Physiology Project, directed by the charismatic marine biologist Terrie Williams. There had been controversy about his removal to California, but his cataracts were becoming worse, and it was hoped that his vision might be saved. Before he left for the mainland, he had been given a Hawaiian name: Ho'ailona. Molokai elder Walter Ritte translated: "special seal with a special purpose." Now he was back, and the excitement was tangible; the aquarium had put out banners announcing his arrival and proclaiming his welcome.[1]

When I got to KP2's area, he was out of the water, and I could see his beautiful silvery fur. His eyes and mouth were closed, and the distance was too great for me to see his lovely whiskers.

Wonderful as it was finally to lay eyes on a real Hawaiian monk seal, there wasn't actually a lot to see. Monk seals spend a lot of time hauled out on warm beaches, sleeping and soaking up the heat of the sun, saving their metabolic energy for the work of deep diving for food. They are marine mammals with a shore life that includes not only sleeping, but also molting, giving birth, and raising pups. Hoʻailona was said to be an outgoing, human-oriented, playful seal, but the chatty little celebrity with his own Facebook page wasn't showing that side of his personality. Instead, he was snoozing.

As we gazed at the peaceful scene, we heard flurries of activity and voices behind us. It seemed that an adult monk seal had just hauled out on Waikiki Beach. We got a few instructions and raced off to see this unexpected event. People who had been walking along the beach were gathering, taking photos, and speaking excitedly in low voices. Already, volunteers had marked the area with orange plastic cones and had taped it off. There were signs cautioning people to be quiet and keep their distance. "Shhhh . . . I'm sleeping," said one attractive notice. Others advised that monk seals are protected both federally and by the state, and that they must not be touched or in any way disturbed. Although a sleeping seal looks something like a 400- to 600-pound slug, seals do have teeth, they do have temperaments, and they do not like to be bothered.

The thrill of seeing a monk seal outside the zoo was interesting in itself. After all, this guy was sleeping just as soundly as Hoʻailona, and I had to examine the question of why the experience was so different. Answers to this question turn on a seeming paradox that is characteristic of many multispecies proximities in this era of extinctions. Although most of the causes of extinction are driven by humans, an ever greater diversity of nonhuman animals are living in crowded cities and suburbs and on beaches. Many of these animals are incorporating into their habitat repertoire areas that humans had thought of as strictly-for-humans. No doubt, the human view of exclusivity has always been an illusion, and of course there has generally been a place for domesticated companion species. But as humans take up more and more space, leaving less and less for others, and managing places in ways that may not be conducive to the well-being of others, new proximities are coming into being, bringing with them new encounters with the mysterious. KP2 could be visited every day of the year except Christmas and Honolulu Marathon Day. Out on the beach, everything was unpredictable, and every haul out was a surprise—a happening.

My research is dedicated to communities that emerge in encounters with animals at the edge of extinction. Such communities coalesce around those in peril. They are contingent, episodic, and imbued with an ethical call. I focus on volunteers and a practice that involves dedication, love, and a profound commitment to the idea that while no death can ultimately be prevented, every encounter can be (and is) experienced as an ethical appeal around which a community can (and will) form.

Those Who Sleep on Beaches

Hawaiian monk seals are members of the Monachus lineage, which evolved in the coastal waters off what is now Turkey and Greece. They were, and remain, a strictly warm-water marine mammal. With time, some of them moved out into the Atlantic and found their way to the warm waters off Africa and to the Caribbean. With even more time, some of them moved from the Atlantic into the Pacific, using a waterway that is now closed. This group found its way to Hawai'i; it became genetically separated from the others about 15 million years ago. The Caribbean monk seal (*Monachus tropicalis*) was last sighted in the 1950s and was declared extinct in 2008. The Mediterranean monk seal (*Monachus monachus*) is right at the edge of extinction, with only about 600 individuals still alive. The Hawaiian monk seal (*Monachus schauinslandi*) has about 1,100 living members. Monk seals are, therefore, one of the rarest families of marine mammals still living (NOAA Fisheries 2013).

These seals inhabit three ecological zones: beaches; shallow coastal waters for pupping, weaning, and foraging; and deeper reef areas for foraging (Kittinger et al. 2012). They are "benthic feeders," meaning that they forage in the ecological zone at the lowest level of the water, including the sediment surface and subsurface. They live on crustaceans, fish, and cephalopods such as octopus. They are apex predators in coral-reef ecosystems, and recent evidence suggest that monk seals have positive effects on island coastal ecosystems through the nutrient transfers involved in churning up sediments (Kittinger et al. 2012; Williams 2012). Aside from humans, their main predators are sharks.

Monk seals show almost no flight response in relation to humans. As long as they are not threatened, they stay put. This lack of fear is part of what made them vulnerable to the commercial slaughter that led them almost to extinction, and the same lack of fear makes them equally vulnerable today to those who wish them harm.[2] It also constitutes a large part of their awesome presence. Entering into such close proximity with wild animals is a great and rare privilege for us humans.[3]

From 15 million years ago until very recently, Hawaiian monk seals knew nothing of humans. Members of our species were late arrivals in the Hawaiian Islands. The great seafaring Polynesians settled in the main Hawaiian Islands (MHI), with their fertile soils and rich coastal environments. There is ongoing debate about the timing of their arrival(s), but it was clearly no later than 1200 C.E. and may have been as early as 300 C.E.. Their migratory path probably brought them from the Marquesas to the Hawaiian Islands, and there was also contact with Tonga (Kirch 2001:80).

The detailed research carried out by John Kittinger and his colleagues (2012) suggests that monk seals left the main islands shortly after Polynesian settlement. By the time of European arrival at the islands, monk seals were living almost exclusively in the uninhabited low islands to the northwest of Kaua'i and Ni'ihau, known as the Northwestern Hawaiian Islands (NWHI) or the Leeward Islands.

Today the monk seal population is declining at a rate of about 4 percent a year. The primary causes of population loss include starvation; entanglement in marine debris, including fish hooks; direct human impacts, such as beach disturbance; loss of haul-out and pupping sites due to beach erosion; diseases; and male aggression toward females. These factors are coupled with low genetic diversity to produce a bleak future (NOAA Fisheries 2013).

As the International Union for Conservation of Nature's Red List of Threatened Species (2016) sums up the situation: "The Hawaiian Monk Seal population is greatly reduced in size from historical levels, has been declining in abundance since at least 1958, and will without question continue to decline for some time into the future. The causes for the decline are only partially understood, have not ceased, and may not be reversible."

Conservation efforts are formally organized by the Fisheries Division of the National Oceanic and Atmospheric Administration (NOAA). Scientific research is coupled with public consultation and education, and with the careful management of individuals in distress. Volunteers are integral to conservation efforts in the main Hawaiian Islands. Without their guardianship, the vast expanse of Hawaiian beaches could not be monitored, sleeping seals could not be protected from curious humans, and injured seals would be unlikely to be found and rescued.

Those Who Watch over the Sleepers

The monk seal volunteers on the island of Kaua'i were generously responsive to my request to carry out research with them. Tim Robinson, projects coordinator of the Kaua'i Monk Seal Watch Program, was particularly welcoming. He is involved, generous, and outgoing, a great communicator, organizer, and activist. He described the group as "an autonomous, all-volunteer group dedicated to the preservation of Hawaiian monk seals through care of them on our beaches, and primarily, through education for our resident and visitor populations" (personal communication).[4] There is another, complementary, group: the Kaua'i Monk Seal Conservation Hui is a volunteer-based project with assistance from state and federal agencies and private organizations. Volunteers report to NOAA's Pacific Island Regional Office and Pacific Islands Fisheries Science Center.[5] Many people work with both groups. A person who knows them well but is not a volunteer described them as "a fantastic cast of passionate characters."

When a seal hauls out, volunteers are called out. They go to the reported site and set up stakes and plastic tape, they put up signs, they make sure the event is reported, and they stay. They are not so much police officers as educators, so while they make sure that people respect the sleeping seal's need to be left alone, they also answer questions about monk seal biology, history, future, and behavior. Although the official literature does not put it this way, volunteers are ambassadors for monk seals. Their commitment to being there is in itself an ethical statement.

In the abstract, the idea of hanging out on a beach all day and keeping an eye on a sleeping seal sounds great! In actual fact, it is both great and not so great. Volunteers often bring with them a chair, water, and food. Seals are champion sleepers, and while volunteers try to relieve each other, a person can end up staying for quite a while. The volunteers have to be knowledgeable, and they need the patience to have the same conversations over and over, day after day. They must be able to discuss conflicting views about monk seals in a manner that doesn't exacerbate conflict. It helps to be a good storyteller as well as a good listener. Along with all these skills, they need a certain tolerance for the bizarre.

Tim told a truly weird story. He arrived at the beach on a call out one day and saw that a woman had approached a large resting seal inside a protective perimeter. She had two small boys with her, and her stated plan was to place the boys atop the seal for photos. Tim calmly but quickly explained that she and the children had to move back outside the perimeter immediately. He followed it up with a bit of education regarding the seal's need for undisturbed rest. The woman said in her own defense that she thought the seal must have been dead!

A site where a pup is born is cordoned off far more extensively, and the watch is particularly crucial. Monk seal mothers, like most animal mothers, are extremely protective of their young, and conflict between humans and seals can easily erupt if humans come too close. In addition, disturbance can break the mother–child bond and wreck a young life before it even gets started. There is a period of about six weeks when mother and pup are inseparable. The pup is totally dependent on the mother for food, and the mother is basically starving as she remains with her pup, providing it with the nourishment that enables it to grow extremely rapidly. Monk seal milk contains up to 65 percent fat. The baby grows while the mother is gradually reduced to skin and bones. Mothers swim in the shallows with their pups, and give them a quick education in being a self-sustaining seal, but at the end of six weeks (or less, depending on the mother's condition), the youngster is left to fend for itself (Williams 2012).

Babies are utterly charming to look at, and people flock to get a glimpse. Keeping visitors at bay, answering questions, and monitoring mother and pup as they move along the shore from beach to beach, the volunteers work on popular beaches throughout the sunlight hours of the day, every day for six weeks or so. They come to know the seals individually, and many people become extremely attached to individuals they have had a lot to do with. At the same time, though, they must preserve distance and not allow the pup to become habituated to humans. KP2's story is a perfect cautionary tale. No one knows why his mother rejected him, but people decided to try to save the lively little pup, who was sucking on stones and starving. He was raised in captivity by humans, and, having been socialized with humans through his early care, his release back into the ocean became an opportunity to find people to interact with. He loved his pink boogie board, and the children of Molokai played with him in the surf. As he grew, he became a potential menace. Seal play gets rough; seals push each other under water, tussling and holding each other down for up to twenty minutes. Efforts to relocate him were unsuccessful; he wanted to be with humans. Once KP2 was back in captivity, the first response was to euthanize him. He was lucky that Williams was able to bring him to her lab, and later he was lucky that a place opened up at the Waikiki Aquarium.

Many of the haul outs on the island of Kaua'i are on well-populated beaches, and so are some of the births. Po'ipū Beach, for example, is fringed with luxury hotels, as well as public parks, and has the reputation of being one of the best beaches in the United States.[6] The year it gained prime recognition was also the year that a pup was born, and the beach had to be closed to humans for a while.

Another volunteer with whom I spent beach time is Kim Steutermann. She is a writer and a wildlife volunteer with a love of the animals of Hawai'i, and with the writing skills to communicate that love. Her articulate account of how she became a monk seal volunteer starts at Po'ipū Beach:

I was thinking about one of your questions as I drove home the other day. You had asked how I had gotten involved, I think. I recalled that I had read an article in the newspaper about a monk seal pup that was born on Poʻipū Beach. I saw photographs of its cuteness, too. The article gave a phone number for volunteers to call and help pup-sit. It was perfect timing—my heart opening up at the same time a hand was reaching out for help. I responded. Who wouldn't! (personal communication)[7]

THOSE WHO SWIM TO THE MAIN ISLANDS

The majority of the small Hawaiian monk seal population lives along the beaches, reefs, and coastal waters of the vast chain of uninhabited atolls of the Northwestern Hawaiian Islands. Federal protection of this region began in 1909, when President Theodore Roosevelt created the Hawaiian Islands Reservation; his main concern was with the seabird nesting areas. In 1940, President Franklin Roosevelt upgraded the protection, and since then various types and levels of protection have been implemented, culminating in the Papahānaumokuākea Marine National Monument—one of the largest marine-conservation areas in the world. The Monument (as it will hereafter be identified) was signed into law by President George W. Bush in 2006 and was expanded by President Barack Obama in 2016 (Office of the Press Secretary 2006, 2016). It encompasses 582,578 square miles of the Pacific Ocean and was intended to protect an array of natural and cultural resources. On July 30, 2010, the Monument was inscribed as a mixed (natural and cultural) World Heritage Site by UNESCO. It includes a number of wildlife refuges and marine protected zones. All fishing is banned, as is any other form of resource extraction.[8]

Hawaiian monk seals are just one of 7,000 marine creatures within the protected area (Office of the Press Secretary 2006). For them, however, protection has not worked as planned. The population within the Monument is declining (now roughly 900 seals), while the small population within the main Hawaiian Islands area is rising (roughly 200 seals). But the decline is happening on a greater scale than the increase. In Williams's (2012) words: "Ten years ago [in 2002] pups born ... within the reserve were almost guaranteed to see their second birthday," but since then, the islands have become "death traps" for pups and juveniles. Williams describes the beaches of these islands vividly: "[T]he emaciated bodies of young seals now littered the shores of the Northwestern Hawaiian Islands."

The story is that since the establishment of the Monument, young seals have been starving to death. Dr. Charles Littnan leads the Hawaiian Monk Seal Research Program at the National Oceanic and Atmospheric Administration. He is based in Honolulu and is, of course, deeply concerned about the disparity between the two regions. In a long and generous interview, he explained that fewer than "one in five pups born survive to adulthood in the northwest Hawaiian Islands. And that has really devastated the population structure."[9] The loss of female pups is particularly grievous, Charles explained: "They're the battery; they're the engine in the population: they're the ones that are constantly recruiting and then having offspring." Speaking specifically of the NWHI, he explained:

> We don't even talk much about recovery at this time. We're in a much shorter time horizon. We're talking about salvaging reproductive potential as the first thing we need to do. . . . We're almost totally focused on females, which I feel terrible about. . . . Because as it is now, if tomorrow we could fix every single problem, we're still going to see a population crash. Because right

now the only thing driving this population is these older females. And they'll go into reproductive senescence or die, more and more every year. . . .

The contrasts with the main Hawaiian Islands are extreme, Charles explained:

[I]t isn't just that they're doing better here. They're doing phenomenally better. A nine-month-old animal looks like a two-year-old in the Monument. Juvenile survival is 60 to 70 percent rather than 15 to 25 precent. An adult female starts pupping in NWHI at seven, eight, ten years of age; in the main islands, we have females that give birth at four. In terms of condition and size and foraging success, they're rock stars down here.

When the Monument was set up, all fishing was banned. The whole zone was left to itself to recover. Starvation has been one of the main results. Charles went on:

You've got these sharks and *ulua* [jack fish]—they have similar diet to the monk seals. So we said—okay, everyone recover. We pull the trigger, and the race for recovery starts. Sharks and fish . . . based on their energetic requirements and their reproductive capacity, they're always going to win out. The spaces that monk seals formerly occupied are taken over by other species.

Along with starvation, there is also outright predation by sharks: "Survival rate from birth to weaning is just about 100 percent. Then a large number disappears or shows up with shark wounds."

In contrast, in the MHI, the ecology is different geologically as well as socially. High islands bring more consistent nutrients into the waters. And sharks and *ulua* are fished out. Very popular items. So even though there's an enormous amount of ocean use by humans, monk seals have found a niche. They're competing with fewer monk seals, with fewer jacks, and fewer sharks.

But coming among humans entails numerous other risks, including dogs, diseases such as toxoplasmosis, and outright murder (Dawson 2010).[10]

This "tale of two regions" (Charles's words) shows the negative trajectory that operates when seals are away from humans, and the positive trajectory that operates among humans. And yet, in the MHI the greatest threat to monk seals is humans. Kim's story about volunteering to pup-sit and asking "Who wouldn't?" is not exactly rhetorical. Not everyone welcomes monk seals.

THOSE WHOSE CULTURES MAY OR MAY NOT INCLUDE MONK SEALS

About seven monk seals are known to have been killed by people in recent years. Only one of the perpetrators has been discovered. His experience seems to have served as a warning to others. A Hawaiian named Daniel Kaneholani, a Kaua'i fisherman, killed a female monk seal, decapitated her, butchered her, and offered sections of meat to bystanders. He was charged with violating the Endangered Species Act. He pleaded guilty and was sentenced to a year in prison (Trask 1998:40). This and other killings speak to a great divide between Native Hawaiians. One side of the divide includes a positive take on monk seals. KP2 is a great example: he was a favorite among the Hawaiians of Molokai, where he hung out. They protested against his being taken away; they named him, blessed him, and never stopped agitating for him to return home. Other people, though, are killing monk seals.

The intrepid Polynesian seafarers brought to Hawai'i a suite of plants and animals previously unknown to the islands. It is not possible now to know how much of the knowledge the people brought with them was transposable to their new home, and how much they needed to learn, but one totally new creature met their gaze: the Hawaiian monk seal.

As island people, they had to ensure that they remained within the carrying capacity of land and sea. Their recognition of the interrelationships among the gods, humans, other animals, plants, climate, geography, weather, land, and sea is expressed in the great creation chant known as the Kumulipo, and in the ancestral-animal power figures known as 'aumakua. Kittinger and his colleagues (2012) conducted an exhaustive survey of the literature on contacts between Hawaiian people and monk seals, and concluded that monk seals were definitely visiting the main Hawaiian Islands when people first arrived. A few archaeological sites show monk seal remains, and the linguistic evidence suggests that people probably hunted monk seals for meat and fur. Monk seals do not figure prominently in either written or oral records, however, leading Kittinger and his co-workers to conclude that monk seals became rare in, perhaps even absent from, the settled islands before the great cultural elaboration of human–animal relationships reached its full flourishing. Monk seals may perhaps appear in the Kumulipo; there is one term, and it is contested (Kittinger et al. 2012:145; Trask 1998:41).

The evidence for contemporary human–monk seal relationships (which may also have old roots) is more compelling. Kittinger and his team met with families that include monk seals among their 'aumakua, and there are some communities that perform ceremonies for monk seals, recognizing the animals as part of the 'ohana, or family. Many of these people were protective of knowledge that is their intellectual property; according to Kittinger (2012), "Respondents have said that the details of such activities are deliberately kept ... secret" (145, 147).

The effect of different histories and knowledges is that some people recognize monk seals as family, while others claim that they are an invasive species, and many subsistence fishermen say that monk seals steal their fish. Controversy is inevitably tangled up in the history of invasion, colonization, dispossession, and ongoing antagonisms between Native Hawaiians and outsiders, particularly the federal government (Trask 2000). These issues are discussed by Thom van Dooren (chapter 6, this volume) and will not be repeated here. The effect is that the fact that NOAA (a federal agency) is in charge of conservation and protection means that it is viewed by many with extreme and enduring suspicion.

Antagonism toward monk seals may be escalating. It is assumed that the murders are done by angry fishermen, and as the most angry antagonists have been Native Hawaiians, it seems reasonable to hypothesize that they may be responsible for the killings. This seems to be the view of Elder Walter Ritte. It was he who gave KP2 his name—Ho'ailona—and he has been a strong and consistent advocate of monk seals. He explained his deep commitment to the lives of Hawaiian creatures:

> Our elders are saying that these seals are not Hawaiian. Our young people are calling these seals an invasive species brought in by government. The seals are now the easy targets of blame for the many ills of our depleting fisheries. We need to stand up for the truth: these seals are not only Hawaiian, but have been here longer than the Hawaiians. These seals are not invasive; they are like the Hawaiian people who are struggling to survive in their own lands. Hawaiians need to see themselves when they see a Hawaiian Monk Seal. How we treat the seals,

is how we can be expected to be treated as Hawaiians in Hawai'i. (quoted in Osher 2011)

The complexities of these social interactions came together in 2009 at the funeral for two monk seals that had been shot that year on the island of Kaua'i. One of the two seals was a five-year-old male, and the other, shockingly, was a pregnant female just about to give birth. The report stated that "she previously had five pups and was an important breeding female and huge loss for the Main Hawaiian Islands monk seal population." Both mother and baby died, so perhaps it is fairer to say that three monk seals were killed.

The ceremony was held on Po'ipū Beach, where the seals' ashes were released into the ocean in a ceremony conducted by Kumu Sabra Kauka, a Kaua'i Native practitioner. State and federal officials took part, as did volunteers and other interested people. Kauka's words echoed those of Ritte: "They [the monk seals] have been on these islands longer than we have; they have been in this ocean longer than we have. They have every bit as much right to live on this earth as we do" (quoted in Zickos 2009).

THOSE WHO REFUSE TO TURN AWAY

Interactions between volunteers and monk seals take place at the nexus of community, ethics, and peril. Ethics are at the heart of it, but the problem of community raises the questions in particularly clear and concise ways.

Traditional ways of thinking about community are based on what we have in common. As Alphonso Lingis (1994) explains in his study of nontraditional communities, a community, conventionally understood, is made up of people who share language, values, and understandings of the world that enable them to sustain their commitment to working together for their common (shared) goals. This type of community is called the "rational community." It is taken for granted that the rational community is a human community.

Clearly, the rational community has multispecies dimensions. Humans form the community, but many of the shared values and goals concern animals with whom the humans interact. KP2, the young seal in the Waikiki Aquarium, is part of a rational community, although this is not a choice he made. A suite of technologies, instrumental logics, and cultural and ethical objectives are bound up in the complex duty of care that human beings have had in relation to him over the course of his life.

KP2's life is situated within a wider rational community of science and technology aimed at conserving and recovering a critically endangered species. While there is no full consensus around methods and means, there is a shared culture of science-based conservation. The scientists who work for NOAA, like Littnan, are part of this rational community. Whether their interactions are with monk seals at large or with monk seals in care, the overarching framework of interaction is set by the rational community.[11]

The volunteers can and do marshal the discourse of the rational community all the time. Their public-education efforts, both formal and informal, work with the shared values and knowledge produced within a rational community. And yet, there is something else, something other or outside, something difficult to talk about because it refuses to be defined by or reduced to the rational community.

People dedicate many hours of every week; are exposed to the elements and to other humans, not all of whom are friendly; and become attached to individual seals they may never see again and with whom they must not become friendly. Sometimes they end up emotionally distraught. They become committed in body, mind, and spirit to creatures with whom they don't share a language, a culture, or a way of life except that they all live near beaches. One of the questions I have been asking volunteers is: "Why do you do this?" Responses to my questions varied in intensity, and all included a desire to make a difference. But for many people, there was a reluctance—indeed, a refusal—to "justify" their commitment, as if their efforts had to be answerable to a conventional logic. Many people said simply, "Because I can." Or, as Kim said, "Who wouldn't?" These brief statements of fact ("I can") or common sense ("Who wouldn't?") refuse justification within the realm of the rational community. That is, a justification within the rational community would bring the discussion back around to what is good for the human community. But they do more: such comments, and refusal to take the matter further, demonstrate a refusal to justify their refusal to justify their actions. In short, they refused all justification.

As a research scholar, I have the honor of taking this refusal seriously—of trying to meet it on its own terms, so to speak. I am seeking therefore to avoid justification. Rather, the aim is to explore a poetics of refusal and to uncover an enhanced vocabulary surrounding refusal. Lingis, Emmanuel Levinas, and Edith Wyschogrod are my philosophical guides.

In the years after World War II, a number of philosophers began addressing the question of ethics and community (for example, Blanchot 1988; Derrida 2005; Nancy 1991).[12] Fascism had shown the terrors of closed communities based on a shared or an imagined monoculture. Questions about traditional communities homed in on boundaries, ethics, and reciprocity. If ethics are shared within the rational community, how can we imagine or understand an imperative toward ethics that arises and commands us from outside the domain of shared values and goals? What of the strangers, the excluded, the refugees, the helpless? What of those who cannot reciprocate?

Alphonso Lingis's book *The Community of Those Who Have Nothing in Common* (1994) draws on the major philosophers in this area and addresses these questions in a form that is close to prose poetry. He is interested in "other" communities—those that do not come into being through what we have in common: "This *other* community is not simply absorbed into the rational community; it recurs, it troubles the rational community" (Lingis 1994:1). Breaking free from that which is shared, Lingis asks what ethics command us in the absence of religion, economic interests, and the solidarity of shared values. In these encounters, meaning arrives mysteriously. We often do not, and may never, understand others with whom we do not share the qualities of the rational community, and yet we recognize that they, too, inhabit worlds of meaning. We acknowledge our shared vulnerability, and it follows that although our ethical responsibilities have no clear rational command, they make claims on us.

This question is connected to the work of Emmanuel Levinas. His life work has been summarized in the single phrase "ethics as first philosophy." Ethics, he argued, again and again, precede the self, arise outside the rational community (the City of Law), and do not form a calculus. His work focuses on the two sides of ethics: the entanglements that bring forth subjectivity, and the refusal to justify or ignore the sufferings of others (Bernasconi 1986; Levinas 1989). One of the terms that Levinas uses to talk about the call is "the face." Whether aural or visual, the other's claim on me arrives to interrupt my self-absorption and to awaken me to my responsibility as a living subject—which is to say, as an ethical subject. As Susan Handelman (1996) explains, "facing is being confronted with, turned toward, facing up to, being judged and being called to by the other" (226).

Levinas situates his analysis in the extremes of suffering and death: "[T]he face is the most basic mode of responsibility. As such . . . the face is the other before death, looking through and exposing death. . . . [T]he face is the other who asks me not to let him die alone, as if to do so were to become an accomplice in his death. Thus the face says to me: 'you shall not kill'" (Levinas and Kearney 1986:23–24).

As Judith Butler (2004:134) reads Levinas, this plea awakens us to the precariousness of the lives of others, and thereby to the precariousness of all life. Levinas's philosophy is supremely relevant to the question of peril, for the heart of ethics is the call from the other. And yet, as is well known, Levinas was equivocal on the question of whether animals are included in ethics in the sense of having a "face," or a commanding ethical presence. A number of excellent recent studies are abolishing Levinas's apparent human-centric vision of the ethical subject (for example, Clark 1999; Llewelyn 1991; Perpich 2012; Steeves 2006). Other scholars are working to include joy (along with suffering) in the analysis of ethics (for example, Mooney 2012).

Levinas (1998) draws a contrast between the self as citizen (that is, a member of the rational community) and the ethical self who experiences "my responsibility for the other . . . , without concern for reciprocity, in my call to help him gratuitously, in the asymmetry of the relation of one to the other" (100–101). One experiences a face, an ethical call, a command from others, and it is only later that someone might ask whether it counts as ethics because the one to whose face one responded was an animal.

Ethical action takes place in a domain of entangled worlds of life and death within which we are exposed to our shared precariousness and express our responsiveness to the vulnerability and suffering of others. Elder Walter Ritte offered a profoundly engaged sense of exposure to monk seals when he described his own moment of revelation in 2006 when he was campaigning to stop development at Laau Point on Molokai. The reporter Jon Mooallem (2013) told the story:

Hundreds of protesters occupied the point for three months, sleeping on the beach. And there, in the quiet, monk seals began to appear on the sand—the first that some protesters had ever seen. Ritte told me that, sleeping side by side—Hawaiians and Hawaiian monk seals—it was just so clear to him: "I was there for survival, and the seals were there for the same reason. I saw myself in the seals."

The final element of this multispecies nexus is the looming threat of extinction. The philosopher Edith Wyschogrod (1990) addresses the question of how Levinas's philosophy relates to community, and argues that human-caused mass death must inevitably affect our understandings of community. I have been arguing that the anthropogenic mass-extinction event now in process is to be understood as another human-caused mass-death event (Rose 2011). This is not to say that genocide and extinction are identical, but rather to pose connections that help us understand both more deeply. As David Clark (1999) has written, it is possible and ethically significant "to think human and animal deaths as capable of illuminating each other in their separate darknesses" (186).

Wyschogrod (1990) uses the term "death event" to identify the nexus wherein the issue of scale—the vast numbers of the dying—is set within the compression of time; that is, it is happening very rapidly. It is a brief but enormous definition that involves human responsibility: large-scale numbers, compressed time, and human agency. She discusses communities in light of the fact that we live in the shadow of, and among, death events, and it is within this world of death events that she analyzes the possibilities for community. She comes to the term "unavowable community" by drawing on Maurice Blanchot (1988), with the aim of reworking his usage to specifically identify the particular community that forms under shadow of what she does not fear to call "apocalypse." The unavowable community has specific characteristics:

1. It "does not lead to any 'communion' or fusion of singular identities, but at the same time affirms both the difference and being-together of singularities."
2. It is involved with death and with the desire to prevent extinction or extermination.
3. It works with unwork (*désoeuvrement* [placing myself at the disposal of another]).

Unwork, or interruptions to everyday rationality, can be thought of, as with Levinas, as a multifaceted refusal: the refusal to justify the suffering of others, the refusal to abandon, the refusal to translate ethics into the rational calculus—which is to say, the refusal to allow the integrity and beauty of ethical call and response to become fodder for rational discourse.

In Wyschogrod's (1990) analysis, the unavowable community, or what I would prefer to call the death-age community, forms in "spaces in the social web where, in the death age, desire for the Other's continued existence can be expressed in discourse and action." It involves "making way for the Other without demanding reciprocity," and therefore it offers a primacy to the other and accepts that there may be no future for those who are being exterminated. The response is not calculated on the probabilities of an outcome. These ethical subjects belong to "an unavowable community, a broken architecture, one without blueprint, of a building without foundation" in this era of mass death. It therefore is "the dream of community in the age of the death event" (175).

THOSE WHO OFFER PRAYERS

The "dream of community" is enacted daily on the beaches of Kaua'i and other islands. Unexpectedly, interestingly, compellingly, a seal arrives. The arrival feels like a message from the ocean, a statement of vulnerability brought onto land by those who live in the ocean but also need to haul out. This moment is a fissure in monoculture, a happening that becomes a recursive attractor that continues to open up *other* communities.

I was back in Kaua'i in May 2012, and there were pups. Kim invited me to join her as she did her guardian stint at Anahola Beach with RK13 and her four-week-old pup. RK13 was a regular on the beaches of the east side of the island, but had always gone somewhere else to bear her pups. This year, she stayed. She was either blind or vision-poor in one eye, and had been bitten by a shark not long before. She went into the quiet waters of the canal to recover, and although she got better, she was not in top condition when her pup was born. There were concerns that she might not be able to continue to feed her baby for much longer, but the expert opinion was that after four weeks the pup would have a good chance of survival.[13] I watched the two of them swimming in the shallow water, and I watched them haul out. RK13's vertebrae and ribs showed through her skin. Gone was the sleek seal look. The pup, in contrast, was shiny and rotund. The pup got yet another feed, and the day seemed incredibly peaceful.

Kim said that she thinks we are attracted to wild animals because we see in them aspects of humanity that we respect. As I understood her words, standing by her on the beach and watching the seals, these encounters offer awareness of ourselves as members of the wider communities of life. They do not tell us that others matter because they are like us. Rather, they tell us that in some difficult-to-articulate way, we and they all matter because we are part of this wide world of life, which is always finding its way into shape, form, and culture through our varied and yet connected ways of being/becoming.

Late in the day, Tim phoned and offered to take me to a beach that was far less frequented and where the pup was only a few days old. We had a long walk on the beach to get to the birth place. Tim is a big man, and he was recovering from broken bones in his foot, and yet he glided through the slippery sand like a creature born to it. I struggled. It was late afternoon on a typically perfect Kaua'i beach. Every cliché was true: the water was turquoise; the sand was white; the sky was blue; gentle breezes carried fresh and tangy air; the sounds of waves and birds were all that could be heard.

We were walking, I thought, through what Levinas calls "the elemental." As Lingis (1994) expresses it:

> We do not relate to the light, the earth, the air, and the warmth only with our individual sensibility and sensuality. We communicate to one another the light our eyes know, the ground that sustains our postures, and the air and the warmth with which we speak. We face one another as condensations of earth, light, air, and warmth and orient one another in the elemental in a primary communication. (122)

At last we came to the mother and pup. The tape barrier was up, and we couldn't go close. We walked away from the ocean, heading back toward some low trees from where we could talk, take photos, and observe with less chance of causing disturbance. That was where we discovered that people had been making prayers.

A prayer in this context is, to quote Michael Leunig (1998), "a small, ancient, wonderful, free-form, do-it-yourself ritual of connection, love and transformation." One of the prayers was set at the base of the tree, on the land side from where the seals were. People had made a circle from sawn logs. They arranged stones in the center and radiated mussel shells outward. They added a feather and pieces of driftwood. They put a piece of driftwood upright in the sand; it was forked, and they put a stone and another shell in the fork, creating a gesture of offering. The artistry lay in the care to show that this was totally nonrandom. At the same time, only found materials were used, creating an offering of local intention. The creator(s) drew on ancient forms—the circle on the ground coupled with the upright column. The prayer was open to embellishment. It gave no sense of being a completed or self-enclosed project. There was always room for another feather, another shell.

The beach was large, and there was another prayer. This one relied on driftwood, shells, and stone, and the creator(s) had brought some fresh flowers to the circle. According to Tim, these arrangements had been put in place shortly after the seal was born. Here, too, there was no sense of a finished work. The prayers were like life—they were gestures in the mode of becoming, and they were offered in the spirit of the face. In Handelman's (1996) terms, the face also involves a verb. "Facing" is not only being confronted with, but also turning toward. An ethics of joy and gratitude, praise and offering, pervades the turning-toward aspect of the face. Perhaps there was also a plea for a safe life for both mother and pup. Just as the physical prayer was open to further embellishment, so were the possible messages of these prayers—if, indeed, they bore messages.

The seals' presence was a blessing; that is, it was a gift to which we have no autonomous right. Such a gift cannot be justified; that is part of what makes it a blessing. It was a gift that we can experience only as an emanation from *outside* (self, expectation, desire). There are recursive effects of blessings—one not only leads to another, but provokes, responds, and remakes others. Indeed, blessings compound, so to walk among them is to walk as a participant in an *other*, or *outside*, mode of becoming.

The more I work with multispecies communities, the more I realize that the big philosophical questions apply to all meaning-making creatures. As humans, we make our own kinds of meaning, and no doubt our meanings are not identical to those of others. But here on the beach, one could see multiple forms of meaning, expression, and creation. Perhaps all of us living creatures make prayers. The mother and pup were involved in their own drama of connection, love, and transformation. Humans responded. These are prayers by and for those who have nothing in common. These are the *other* prayers that enact through gratitude and commitment the *outside* significance of life—that is, life in the mode of blessing.

ACKNOWLEDGMENTS

Special thanks to all the people in Hawai'i who are involved with monk seals and who shared their time and thoughts as I carried out this research. Many of the ideas in this paper first took shape in conversation with members of the Extinction Studies Working Group, and I am deeply grateful for all the long, deep conversations. The Australian Research Council funded Thom van Dooren and me to carry out research "at the edge of extinction" (DP110102886).

NOTES

1. This brief account is summarized from Williams (2012), which was published a few months after KP2 returned to Hawai'i.

2. Figures are lacking, but one example speaks volumes: "In 1824 a sealing expedition by the brig 'Aiona' was thought to have taken the last monk seal; but after a 'sealing and exploring voyage' to the Leeward Islands between April 26 and August 7, 1859, Capt. N. C. Brooks of the bark 'Gambia' returned to Honolulu having 'on board 240 bbls., seal oil, 1,500 skins'" (Kenyon and Rice 1959:215).

3. Thom van Dooren (2014:36) discusses this beautifully in relation to albatrosses who similarly do not have a flight response to humans.

4. For information, see Kauaʻi Monk Seal Watch Program, http://www.kauaimonkseal.com/Home.html.

5. For information, see NOAA Fisheries, "Protected Resources," National Oceanic and Atmospheric Administration, http://www.fpir.noaa.gov/PRD/prd_volunteer_opps.html. *Hui* means "group" or "association."

6. In 2001, Poʻipū Beach was ranked as the best beach in America by Stephen Leatherman of Florida International University, also known as "Dr. Beach." This fact was offered to me by numerous people, and the details are also reported in "Poʻipū Beach Park," Wikipedia, https://en.wikipedia.org/wiki/Poipu_Beach_Park. The beach was also named "America's Best Beach" by the Travel Channel, which ranked it "top among the 10 'best' beaches selected nationwide" (Poipu Beach, http://poipubeach.org/beaches/poipu-beach/).

7. For a sample of Kim Steutermann's writings, see http://www.outrigger.com/explore/blog?page=29. Her website is https://kimsrogers.wordpress.com.

8. For information, see Papahānaumokuākea Marine National Monument, National Oceanic and Atmospheric Administration, http://www.papahanaumokuakea.gov/about/.

9. This and all subsequent quotes, unless otherwise noted, are taken from Charles Littnam, interview with author.

10. As Williams (2012) explains, monk seals are "immuno-naïve" (93) because of their having been isolated for millennia.

11. I am well aware that much that is accomplished by bureaucracies is not rational in the usual sense of logic and analysis. Many of the people who work in these bureaucracies are also under no illusions about the gaps between reason, rational expediency, and, in many instances, dysfunction.

12. Kevin Hart (2013) notes that Derrida almost never used the term "community."

13. Summarized from conversations with both Kim Steutermann and Tim Robinson. See also Azambuja (2012).

REFERENCES

Azambuja, Leo. 2012. "Injured Monk Seal Recovering." *Garden Island* (Kauai, Hawaii), January 3. http://thegardenisland.com/news/local /injured-monk-seal-recovering/article_b57c49fc-36b4-11e1-a2ce-0019bb2963f4.html?mode=story#ixzz1iZ1IXcAf.

Bernasconi, Robert. 1986. "Levinas and Derrida: The Question of the Closure of Metaphysics." In *Face to Face with Levinas*, edited by Richard Cohen, 181–202. Albany: State University of New York Press.

Blanchot, Maurice. 1988. *The Unavowable Community*. Translated by Pierre Joris. Barrytown, N.Y.: Station Hill Press.

Butler, Judith. 2004. *Precarious Life: The Powers of Mourning and Violence*. London: Verso.

Clark, David. 1999. "On Being 'the Last Kantian in Nazi Germany': Dwelling with Animals After Levinas." In *Animal Acts: Configuring the Human in Western History*, edited by Jennifer Ham and Matthew Senior, 165–198. New York: Routledge.

Dawson, Teresa. 2010. "A New Threat to Hawaiian Monk Seals: Cat Parasite Carried by Runoff, Sewage." *Environmental Health News*, December 7. http://www.environmentalhealthnews.org/ehs/news /hawaiian-monk-seals.

Derrida, Jacques. 2005. *On Touching—Jean-Luc Nancy*. Translated by Christine Irizarry. Stanford, Calif.: Stanford University Press.

Handelman, Susan. 1996. "The 'Torah' of Criticism and the Criticism of Torah: Recuperating the Pedagogical Moment." In *Interpreting Judaism in a Postmodern Age*, edited by Steven Kepnes, 221–239. New York: New York University Press.

Hart, Kevin. 2013. "Four or Five Words in Derrida." In *Living Together: Jacques Derrida's Communities of Violence and Peace*, edited by Elisabeth Weber, 168–187. New York: Fordham University Press.

International Union for Conservation of Nature. 2016. "*Monachus schauinslandi*." The IUCN Red List of Threatened Species. http:// www.iucnredlist.org/details/13654/0.

Kenyon, Karl W., and Dale W. Rice. 1959. "Life History of the Hawaiian Monk Seal." *Pacific Science* 13, no. 3:215–252.

Kirch, Patrick Vinton, and Roger C. Green. 2001. *Hawaiki, Ancestral Polynesia: An Essay in Historical Anthropology.* Cambridge: Cambridge University Press.

Kittinger, John, Trisann Bambico, Trisha Watson, and Edward Glazier. 2012. "Sociocultural Significance of the Endangered Hawaiian Monk Seal and the Human Dimensions of Conservation Planning." *Endangered Species Research* 17:139–156.

Leunig, Michael. 1998. *A Common Prayer.* Melbourne: HarperCollins.

Levinas, Emmanuel. 1989. "Ethics as First Philosophy." In *The Levinas Reader*, edited by Sean Hand, 75–87. Malden, Mass.: Blackwell.

——. 1998. "Useless Suffering." In *Entre-Nous: Thinking-of-the-Other.* Translated by Michael B. Smith and Barbara Harshav. New York: Columbia University Press.

Levinas, Emmanuel, and Richard Kearney. 1986. "Dialogue with Emmanuel Levinas." In *Face to Face with Levinas*, edited by Richard Cohen, 13–33. Albany: State University of New York Press.

Lingis, Alphonso. 1994. *The Community of Those Who Have Nothing in Common.* Bloomington: Indiana University Press.

Llewelyn, John. 1991. "Am I Obsessed by Bobby? (Humanism of the Other Animal)." In *Re-Reading Levinas*, edited by Robert Bernasconi and Simon Critchley, 234–245. London: Athlone Press.

Mooallem, Jon. 2013. "Who Would Kill a Monk Seal?" *New York Times*, May 8, http://www.nytimes.com/2013/05/12/magazine/who-would-kill-a-monk-seal.html accessed 16/5/13.

Mooney, Edward, and Lyman Mower. 2012. "Witness to the Face of a River: Thinking with Levinas and Thoreau." In *Facing Nature: Levinas and Environmental Thought*, edited by William Edelglass, James Hatley, and Christian Diehm, 279–299. Pittsburgh: Duquesne University Press.

Nancy, Jean-Luc. 1991. *The Inoperative Community*. Translated by Peter Connor, Lisa Garbus, Michael Holland, and Simona Sawhney. Minneapolis: University of Minnesota Press.

NOAA Fisheries. 2013. "Hawaiian Monk Seal (*Monachus schauinslandi*)." National Oceanic and Atmospheric Administration. http://www .fisheries.noaa.gov/pr/species/mammals/seals/hawaiian-monk -seal.html.

Office of the Press Secretary. 2006. "Fact Sheet: The Northwestern Hawaiian Islands Marine National Monument: A Commitment to Good Stewardship of Our Natural Resources." Press release, June 15. White House. https://georgewbush-whitehouse.archives.gov /news/releases/2006/06/20060615-9.html.

——. 2016. "Fact Sheet: President Obama to Create the World's Largest Marine Protected Area." Press release, August 26. White House. https://www.whitehouse.gov/the-press-office/2016/08/26 /fact-sheet-president-obama-create-worlds-largest-marine-pro tected-area.

Osher, Wendy. 2011. "Moloka'i Monk Seal Deaths Deemed Suspicious." *MauiNow*, December 23. http://mauinow.com/2011/12/23/moloka %E2%80%98i-monk-seal-deaths-under-investigation/.

Perpich, Diane. 2012. "Scarce Resources? Levinas, Animals, and the Environment." In *Facing Nature: Levinas and Environmental Thought*, edited by William Edelglass, James Hatley, and Christian Diehm, 67–94. Pittsburgh: Duquesne University Press.

Rose, Deborah Bird. 2011. *Wild Dog Dreaming: Love and Extinction*. Charlottesville: University of Virginia Press.

——. 2013. "In the Shadow of All This Death." In *Animal Death*, edited by Jay Johnston and Fiona Probyn-Rapsey, 1–20. Sydney: Sydney University Press.

Steeves, H. Peter. 2006. "Lost Dog." In *The Things Themselves: Phenomenology and the Return to the Everyday*, 49–63. Albany: State University of New York Press.

Trask, Haunani-Kay. 1998. "The Dog That Runs in the Rough Seas." In *Intimate Nature: The Bond Between Women and Animals*, edited by Linda Hogan, Deena Metzger, and Brenda Peterson, 37–45. New York: Fawcett.

——. 2000. "The Struggle for Hawaiian Sovereignty—Introduction." In "Problems in Paradise: Sovereignty in the Pacific." Special issue, *Cultural Survival Quarterly* 24, no. 1.

van Dooren, Thom. 2014. *Flight Ways: Life and Loss at the Edge of Extinction*. New York: Columbia University Press.

Williams, Terrie. 2012. *The Odyssey of KP2: An Orphan Seal, a Marine Biologist, and the Fight to Save a Species*. New York: Penguin Press.

Wyschogrod, Edith. 1990. "Man-Made Mass Death: Shifting Concepts of Community." *Journal of the American Academy of Religion* 58, no. 2:165–176.

Zickos, Coco. 2009. "Seals Get Hawaiian Funeral." *Garden Island* (Kauai, Hawaii), June 19. http://thegardenisland.com/news/local/seals-get -hawaiian-funeral/article_f0564eff-7a90-5a5e-ab9d-dcd 4583903a4.html#ixzz1ibo4dwYX.

Kate Foster, *Dedication to PTT ID 56280*, 2015. (© Kate Foster)

5. ENCOUNTERING LEATHERBACKS IN MULTISPECIES KNOTS OF TIME

MICHELLE BASTIAN

Exploring what it might mean to write in a time of extinctions, Deborah Bird Rose (2013) proposes that one must take seriously the way that "living beings call and respond; ethics are situated in bodies, in time, in place and necessarily, in encounter" (6). To write about extinction ethically, she suggests, is not to write in the abstract, but to understand how the confluence of forces making up this process might connect with the "present temporalities, localities, and relationalities of our actual lives" (6). In what follows, I offer my own attempt to take these words to heart and to write in response. My focus is the threatened extinction of the leatherback turtle, and how to understand this as something more than a crisis happening in a wide blue elsewhere.

Of course, one of the difficulties of attempting this is that leatherbacks rarely enter into the great majority of people's lives with any directness. When I first encountered them, it was as a potential object of research. I had heard of them, but just barely, and I had certainly never seen one. It seemed that in all likelihood I never would, unless I somehow managed a trip to Costa Rica, Trinidad, or Florida. So, as you will read for yourself shortly, building connections that might embed leatherbacks and me in "shared, or partially shared, lifeworlds," as Rose (2013:5) suggests, ended up taking a circuitous route by way of clocks, filing cabinets, conference deadlines, journal articles, fellow commuters, YouTube videos, and a walk along Edinburgh's Water of Leith. While most of these elements will become clearer as this chapter unfolds, an explanation for why clocks appear in this list is in order.

I've come to be fascinated by what it is that clocks do, and particularly what they *might* do (Bastian 2012, forthcoming). Long detested as the device that surveils, enforces, admonishes, ignores, and reduces, the clock nonetheless offers a fascinating window into some of the ways that processes of connection are facilitated and managed. By offering a mesh that encompasses the globe—in the form of Coordinated Universal Time (UTC)—clocks suggest that everything is, in principle, able to be connected with everything else. They promise that we are all together in the same moment, in the same ticking of the second hand. Increasing accuracy has been crucial to this process. Temperature, humidity, movement, sudden shocks, gravitational effects, electromagnetic effects, and more call materials to respond, and when they do the clocks made from them become less accurate and less reliable. And so the process of creating this mesh of connection has been marked by the search for materials and devices that are less and less likely to respond to the environmental conditions around them (Bastian 2014; Mann 2014).

Telling time in a time of extinctions poses different problems. A point highlighted by Rose (2012) in her account of the ethics of multispecies temporality. Focusing on sequence and synchrony, rather than on accuracy and universality, she tells a story of the coevolved relationships between flying foxes and eucalyptus trees. Rose describes the way that synchronies between species—where flowering eucalypts offer sustenance to the migrating foxes, who, in turn, pollinate the trees—sustain each of them through sequences of generational time. Neither sequence nor synchrony happen automatically, but both are embodied achievements. The flying foxes and the trees must find each other, and at the right times. As Thom van Dooren (2014b) writes, sequences depend on "real embodied generations—ancestors and descendants—in rich but imperfect relationships of inheritance, nourishment, and care" (27–29). Neither do synchronies and sequences occur in isolation; rather, multitudes of them bring together food and fed, pollinator and pollinated, traveler and medium traveled. In the case of flying foxes and forests, however, as both of their populations decrease, these "multispecies knots of time" are fraying, threatening the functional extinction that precedes the actual (Rose 2012:138). As this volume shows, these are only one set of knots among many. Thus Rose's proposal is that, with the loss of these relationships in a time of extinctions, time itself is being unmade.

What, then, of the clocks that so often chart our way through relationality? Why summon them here to guide us into a story of turtles? Only a hunch and a hope that they might one day work differently. All clocks are not the same, after all. Within research on circadian rhythms, for example, the environmental conditions that promote responses from body clocks do not threaten time's accuracy, as they do for their namesakes.[1] Instead, elements of daily life that affect embodied time—such as light, temperature, eating, and socializing—are known as *Zeitgebers*, or "time givers" (Pittend-righ 1981). For these clocks, time cannot exist in isolation but is given in relationship. Here accuracy is not about keeping to a regular disinterested beat, but adjusting to the shifting cycles that make life possible. At the heart of this chapter, then, is the question of what happens when the experiences of leatherbacks are drawn into everyday experiences and further, once there, what kind of "time givers" might they prove to be?

Tuesday, February 8, 2005

8:01 P.M.

My partner and I have just moved to Sydney and have been in our new flat for only four days. The four years of my doctorate stretch out unknown before me. We are just arriving home from getting the groceries, and we pause on the side entry to look out over Coogee Beach. It is a new moon, and the ocean is dark. We wonder, like we always will, what might be happening out there, over the water.

Nine years later, I find out. Trying to retrace where I was on that date at that time, I shuffle through filing cabinets, flip through appointment diaries, and consult old rental agreements, as well as weekday and moon-phase calculators.[2] Playing the game of "where were you when this happened?" I collude with the clock and its promise of an all-encompassing time. It offers me a retroactive synchrony that connects that place with another, allowing a leatherback to weave its way into my life.

Because at the same moment that we are standing there, out over that water, on the other side of the Pacific, on another beach, leatherback turtles are hauling themselves up onto land (Shillinger et al. 2010:222).

Over in Playa Grande, Costa Rica, it is 3:01 A.M. The local time is different, but the darkness of the ocean remains, the new moon shared across the globe long before international timekeeping agreements. Since October, female leatherbacks have been congregating offshore, making multiple trips to the beach to nest. Laying between October and February links their reproductive cycles to the cycles of the ocean, with large seasonal eddies helping to pull hatchlings out to sea when they eventually venture forth (Shillinger et al. 2012:1).[3] Like my partner and me, the turtles have been watching the moon, often preferring to wait for a dark night like tonight before making the risky trip onto land.

Aware of these cycles and hoping to play some role in their continuation, human researchers have congregated on the beach. They are responding to the threat of the leatherback's imminent extinction. This threat had been announced five (long) years before (Spotila et al. 2000). And this particular population of eastern Pacific leatherbacks has declined by up to 90 percent in twenty (short) years (Shillinger et al. 2010:215). In other places, they have disappeared entirely. As James Spotila and his colleagues (2000) note, "Leatherbacks had disappeared from India before 1930, declined to near zero in Sri Lanka by 1994, and fallen from thousands to two in Malaysia by 1994" (529). On Playa Grande, there is still hope that the population will recover. This beach is one of their key nesting sites and has been designated as a national marine park since 1991. Egg harvesting has been reduced, and hatcheries have been created to save threatened nests.

But a focus on the short life stages spent on land can only do so much. The intensification of open-sea fisheries in the eastern Pacific, including the use of longlines and gill nets, has had a swift and massive impact. Drawn to the same productive upwellings out to sea, a new synchrony between humans and turtles—one in search of swordfish and the other, jellyfish—has created the conditions for the extinction of a species. Roland Brañas, a local fisherman from Chile, remembers that "before ever using nets, leatherbacks were extremely odd, some fishermen perhaps couldn't even tell them apart from other sea turtles" (Arauz 1999:14). During the late 1980s and early 1990s, however, he estimated that each boat in his area would catch around thirty leatherbacks a year. As early as the mid- to late 1990s, Brañas no longer heard of them, and the leatherback had again become rare (Arauz 1999:14–15). Overall estimates suggest that fisheries "killed at least 1,500 female leatherbacks per year in the Pacific during the 1990s" (Spotila et al. 2000:530). Both before and after, an encounter with a leatherback at sea was a curiosity, but while in one moment this rarity supported the continuation of life, in the other it signaled a decimation.

And so up on the beach, researchers are attaching satellite trackers to the leatherbacks. The hope is that if they can track where the turtles go once they finish nesting, perhaps they can help undo this deadly sharing of time. Inspired by the TurtleWatch mapping tool, which has helped longline fishers in Hawai'i avoid dangerous interactions with loggerhead turtles,[4] the researchers here plan their own "clocks," ones that draw on growing knowledge of how turtles move and their ways of reading the ocean as they search out their prey (Shillinger et al. 2008:1414). If they can discover a pattern, they will be able to suggest that "dynamic time-area closures" be put in place in the southeastern Pacific, like those used by the Hawaiian TurtleWatch, where the boundaries of conservation zones are set, and reset, based on current conditions and their likelihood of attracting turtles, rather than on static geographical borders (Shillinger et al. 2008:1409).[5]

Importantly, for George Shillinger and his colleagues, this time(and space)-telling device is not being built in the service of connecting across distance, but rather to separate human from turtle. Their turtle watch fosters asynchrony, using specific, embodied understandings of time to deliberately disconnect (Shillinger et al. 2011:286; see also Benson et al. 2011). As Brañas's story suggests, the knots of time that support life may also have to be read in reverse—for the patterns of de-synchrony, dis-coordination, and disconnection, which may have been just as important for sustaining generational sequences of leatherbacks and others as the synchronies that Rose (2012) highlights.

When the time is right, the tagging begins; so far this season, four turtles have been added to the project's growing list of tracked animals. Tonight, there are two more: PTT ID 56268 and PTT ID 56280 (Shillinger et al. 2010:222).[6] PTT ID 56280 was first identified in the 1994/1995 season and has been seen back at Playa Grande four times since then. This year, she first hauled up on the night of January 17, and her last return will be on March 1, when she will end her time of inter-nesting and head back to her foraging grounds. As she travels, her tag will send intermittent data to the Argos GPS tracking system, with the researchers following closely all the while.

In the months and years after her tagging, turtle PTT ID 56280 starts to stand out in the analyses of this particular data set. I first came across her in July 2012, learning something of the poetry buried in the strict form of scientific papers. While all the other turtles from Playa Grande headed out into the Pacific toward the Galápagos Islands, Shillinger and his team (2008) reported that "a single turtle in this study (tag ID 56280, tagged during 2005) occupied exclusively nearshore foraging habitats along the coast of Central America throughout the entirety of its tracking duration (562 d)" (1411). Such a matter-of-fact tone, yet in the midst of all the graphs and statistics, her journey insisted on telling its own story. Its implications rushed out and ahead and around. Why was she the only one? What had happened to all the others? How many might there once have been? Did she notice their absence?

I am not the only one to wonder. I trace hypotheses through other papers that mention her. Given the large number of leatherbacks caught in fisheries off Peru and Chile, turtle PTT ID 56280 might represent one of the few remaining "coastal" leatherbacks from a population that is on the very edge of localized extinction (Saba et al. 2008:657). Given the diversity of migration paths utilized by other leatherback populations, it seems unusual that eastern Pacific leatherbacks would have only one (Shillinger et al. 2008:1411). Indeed, these coastal leatherbacks may have been one of the most successful populations in the eastern Pacific, with their foraging areas being more productive and, importantly, more predictable than the open seas of the southeastern Pacific (Saba et al. 2008:657). Eating well requires a particular confluence of temporalities. Being able to predict when and where food will arrive allows a more efficient use of your own resources. PTT 56280 herself was one of the largest tagged in this particular data set; she had larger than average clutches and reached areas where she could forage much sooner than other tracked turtles (Bailey, Fossette, et al. 2012).

Even so, when a particular population has dropped by 90 percent or more in such a short time, and there have been little to no systematic records kept, how can such speculations be answered? As Karen Bjorndal and Alan Bolten (2003) argue, "Many sea turtle populations of today are ghosts . . . of past populations" (16). Who knows how many ghosts may be accompanying PTT 56280 on her solitary journey. Excitement over "soaring" numbers of nesting sites in Puerto Rico, where more than 1,700 were seen in the first half of the 2014 season (EFE 2014), pales in comparison with stories of thousands of nests in a single night on Playa Grande. But anecdotes like these are few, often forgotten or misremembered, and they don't translate easily into the particular language of scientific practice (Pauly 1995).

In an ocean thick with hauntings, what kind of clock could set things to right? Shillinger and his colleagues (2011) hope that their complex of asynchronies will, supporting new forms of reckoning in this time of leatherback extirpations. If they can get it running, their clock promises to remove (some) dangers for (some) eastern Pacific turtles. But to do so, their research must be translated into politically viable objectives. The press release that does some of this translation work shows complexities already being smoothed over. In it, the unusual (and improbable) discovery that eastern Pacific leatherbacks (or at least those who remain) "consistently follow a relatively narrow corridor out into the sea, past the Galapagos Islands and across the equator to an area in the South Pacific" is heralded as "the key to the leatherbacks' salvation" (Stanford University 2008). As the title of the research paper describing this discovery suggests, "persistent leatherback turtle migrations present opportunities for conservation" (Shillinger et al. 2008).

What a relief to find a consistency within the context of so much loss, a stillness inside the chaos. PTT 56280 appears as an interesting oddity, and the implications of her existence are left to future studies. Here and now, the promise of a dependable and limited migration corridor is pragmatically given priority. It offers the kind of time that is most needed for knitting together the range of national and international bodies that might support its continuation. Easier to negotiate with the time of the living, perhaps, than with the time of ghosts.

Yet the ghosts refuse to be banished. I hear them quite close by. The steady tick that offers (on occasion) a sense of safety, of predictability and calculability, has been transposed into an eerie clattering.

TUESDAY, JULY 9, 2013

2:17 A.M.

I am far from home, cold, tired, and anxious. My doctorate is long finished, and home is now on the other side of the world in Edinburgh. There it is 5:17 P.M. Once again, the clock connects me, weaving distant others into the present. Knowing the time, I can guess that my partner will be getting home from work soon. Other homes in Edinburgh will be filling up with returning occupants.

Putting down their bags, making a hot drink, thinking "what shall we have for tea?"

I have found my way back to Sydney, but to a time that is out of sync. I sit at my desk, still awake. Almost everyone else is fast asleep. The traffic has lulled, and the birds are quiet. A clock ticks, steady when I listen for it, but when my attention breaks and I focus back on my screen, it seems to move faster. Suddenly it is 2:54 A.M.

Tick, tick, tick.

Hurry up, hurry up, hurry up.

You'll be late; you won't have any sleep; you don't have time.

I have traveled here to talk about leatherbacks at a conference on animal studies, but I still haven't written my paper and I'm presenting it tomorrow. All the times when I could have done something, could have acted, could have been one of those well-timed and responsible academics, weigh heavily. The consistent and persistent version of myself is yet to be realized, and, as usual, I have procrastinated and put it all off.

I think back to earlier in the evening, when I spent that extra half hour at the opening of the conference exhibition.[7] Or the half hour afterward, waiting at my favorite vegetarian place for steamed dumplings. After I'd spent twenty minutes deciding what to get, enjoying the luxury of so many options.

Despite my pleas, the clock is implacable and won't return the time I've lost.

I am not alone in this time, though. Others will still be awake, working on their presentations. All of us shrugging our shoulders at the gallery, colluding with one another to put it off a little longer. "It'll get written sometime." Now here we are, in this time outside of time, a synchrony of untimeliness.

Still the clock ticks.

Tick, tick, tick.

In a session yesterday, we wondered over the meaning of a magpie's song, but we never doubt the meaningfulness of this monotonous tick. Too early, too late, so bored, can't wait. The clock sings to us in its own way. It tells us stories of late trains, of exams, of cinema screenings, of job interviews, of grant deadlines (not a millisecond after 4:00 P.M.).

We are told there is only one clock time, a rigid mechanical process that is "unaffected by context and seasons" (Adam 1998:70). Your watch might be two minutes faster than mine, but that is not because it is like the magpie, calling us to see it as a unique, creative creature. It is simply because it is wrong. It has the wrong time.

But they once said that the Pied Butcherbird sings only by instinct. Not convinced, a musician and researcher has spent years listened attentively to their song (Taylor 2008). The uneconomic practice of simply spending-time-with produces the "sharp ear" that could move beyond hearing only mimicry and repetition (Taylor 2013). Individuals become distinct, and their song now rings clearly as the voice of a unique being exploring its world.

Maybe we haven't been listening to clocks attentively enough either. Maybe we've just been poring over their bones, clacking them together—clack, clack, clack—and thinking that we know all there is to know about time and the rhythms that bind beings together.

Our clocks promise that they can keep us coordinated, that if we plan sensibly, all will take place as it should. The lure of persistent consistency still guiding understandings of how best to act and respond in the face of existential threats. But what if, in this time of extinctions, our hours are muddled, our dates disoriented, our counting confused?

I hold the clock's bones in my hands, wondering how they might work differently. Time is not what it once was, and all around rhythms are shifting and transforming.

It's now 3:41 A.M.

The bones have started growing flesh.[8]

The cold predawn has me lying on the carpet and soaking up the radiant heat from the under-floor system. I am reading scientific articles, as precise and dry as ever. Despite the authors' best intentions, I evade the longlines of scientific rationalism. Instead, the dark carpet in my room morphs into black sand. I am on another beach, Tortuguero, on Costa Rica's Atlantic coast. In this place, the time is "peak leatherback nesting season" (Veríssimoa et al. 2012).

A jaguar ventures out of the cover of trees. She, too, is under the close eye of human researchers, tagged and tracked, as part of an attempt to halt the fast downward trend of jaguar populations across the Americas (Carrillo, Fuller, and Saenz 2009). Tonight, she is hungry and is seeking unusual prey—sea turtles. Until recently, she had no need to hunt this well-protected quarry. The forest was large and held many options for her. She could take peccaries, monkeys, agouti, or many different kinds of birds or fish. But the forest grows smaller, and so do her choices.

Then, if you had asked her the time, she might have told you about following white-lipped peccaries from uplands to coastal forest to swamps, as the wet season turned to the dry (Carrillo, Saenz, and Fuller 2002). Or of hunting during mornings and late afternoons when the peccaries were out foraging for their food, both peccary and jaguar resting during the mid-day heat (Carrillo, Fuller, and Saenz 2009). But with their numbers dwindling, she has become attentive to a new rhythm. This clock does not signal through the shifting scents of ripening fruits, but by the sound of bodies dragging themselves out of the ocean.

Ever the opportunist, she has begun to forge new relationships of predator and prey. To do this, she has also had to forge a new time. She has noticed that the turtles arrive with the new moon and adjusts her sleep to coordinate with them (Carrillo, Fuller, and Saenz 2009:565). Synchrony is made flesh in her desire to sustain her own life. The beach now holds jaguar and turtle in a fraught and fragile shared moment.

Usually she finds green turtles digging out their nests. But she is early, the time not quite yet "peak green turtle nesting season." Tonight, something huge and unexpected has hauled itself out of the water. Although it does not look very much at all like the others, she is still able to attack the leatherback's vulnerable flippers and neck. Perhaps next year, she will show a cub how to take advantage of the unprotected flesh; the cub, in her turn, might bring her own cubs to feed on this becoming-familiar prey. A new synchrony in the present extending out toward new futures.

I later read that this is indeed what researchers in Tortuguero have found, suggesting that the taking and sharing of turtle carcasses may be "the result of a locally learned behavior, passed down several generations, which [has] now become prevalent across the jaguars living in the area" (Guilder et al. 2015:71). Encroaching agricultural activities, including banana and pineapple plantations, as well as illegal hunting in the national park, have pressed jaguars into finding new food sources.

But learning to kill a turtle also involves learning its temporalities and spatialities, being in the right place at the right time, hoping for prey that is both available and reliably so (Arroyo-Arce, Guilder, and Salom-Pérez 2014:1455). Like my clock, the new moon promises the jaguar that if she keeps to the right rhythms, all will take place as it should. But how many turtles will survive and return next year? How much habitat will she have left? And what are conservationists to do when one endangered species starts eating another (Veríssimoa et al. 2012)? While there is evidence that, unlike the eastern Pacific leatherbacks, the turtles tied to these Atlantic coasts are increasing in number (for example, Stewart et al. 2010), the Tortuguero population is still decreasing (Gordon and Harrison 2012).

Created year after year, synchronies become a sequence through generational time. Or at least they used to. Since the last great extinction event, the tangle and weave of embodied time has grown increasingly ornate and precise, but here in the midst of another such event time is becoming threadbare. The forests, the peccaries, the jaguars, the leatherbacks—all are under threat. They will shift and adapt, seeking out gaps and openings that might remake the rhythms that support their lives.

And so time ends and time begins, with different consequences rippling out for each of those bound up in the knot.

During the peak green turtle nesting season, jaguars often leave much of the carcass untouched. Abundance means that they don't have to take the time to laboriously claw out the hard-to-access meat (Guilder et al. 2015). Not seeing the point in letting the turtle meat rot, local people propose to the park management that they be given rights to the fresh carcasses (Campbell 2007:322). Unlike the jaguar, they draw on centuries-long histories of eating sea turtles, including leatherback. But this request is denied. In a time of extinction, a human encounter with a turtle is not supposed to be about food, but about tourism and research. Nesting season closes public beaches for locals, but opens them for foreign visitors taking advantage of gaps in their own time to "see the turtles" (Campbell and Smith 2005:179). The new temporalities that press the jaguar and turtle into connection, disrupt and disconnect others.

Journal articles are scattered all around me now, here on the warm carpet. I reach for one at random and am swept out even farther, all the way to the other side of the Atlantic (Witt et al. 2007). With the time now "jellyfish season," the leatherbacks have shifted from prey to predator. They have been searching out the optimum conditions for jellyfish blooms. Conservation scientists are slowly piecing together the multiple factors that each turtle attends to in order to be in the right place at the right time. Underlying search rules begin piling up: "Ekman upwellings," chlorophyll-a levels, sea-surface temperature, eddies, swells, choppiness, and currents (Bailey, Benson, et al. 2012; Benson et al. 2011; Hays 2008; Heaslip et al. 2012; James et al. 2005; Witt et al. 2007). Unlike our own context-insulated clocks, leatherbacks' modes of coordination trace intersections between a range of dynamic environmental conditions.[9] Constructing clocks of their own, but so different from the one ticking here in my room.

Once the blooms are found, the turtles can settle into methodically eating their prey, the sheer abundance of jellyfish allowing them almost to graze (Heaslip et al. 2012:6). Like the jaguar, their daily rhythms track those of their prey, rising to the surface at night and sinking down during the day (Witt et al. 2007:237). While the jaguar's body has not yet invented an efficient way of getting into a large turtle's carapace, the leatherback's has had the time it needed to find solutions to its own problems. Jelly after jelly gets pulled into its spiny throat. Known for its immense size, a leatherback is nonetheless capable of eating its own body weight in a day.

The turtles are off the coast of Ireland, feasting on blooms of barrel jellyfish 4 square miles wide (Houghton et al. 2006:1967). Until the publication of Jonathan Houghton's paper, these consistent aggregations of jellyfish in the northeastern Atlantic were unknown to science. Indeed, in the articles scattered around me, marine biologists and ecologists have been lamenting how little is known about jellyfish: when they bloom, how, why, or where. Until very recently, there has been no funding for research and no interest from policy makers in learning more about them (Doyle et al. 2008; Hay 2006; Houghton et al. 2006; Richardson et al. 2009).[10] They are a form of life that humans seem to feel no need to synchronize with. Leatherbacks, though, draw on sequences 110 million years long, knowing where to be, and when, in order to create the beneficial synchronies that make futures.

Unlike with the jaguar, however, the fear is not that the leatherback's prey is decreasing, but that it may be exploding exponentially (Richardson et al. 2009).[11] Many of the human activities that have contributed to the swift reduction in leatherback populations might, perversely, be turning the oceans into a perfect habitat for jellyfish (Purcell, Uye, and Lo 2007). Where once there were stories of fish being so abundant that, during salmon runs, rivers might contain more fish than water (for example, Roberts 2007:45–57), now jellyfish are shutting down tourist resorts, killing fish farms, and blocking intake valves of nuclear power plants (Danigelis 2013). Clearing them from the Orot Rabin coal-fired power station in Israel in 2011 required diggers and shipping containers (Kwek 2011). As with fears of the rise of superweeds on land, abundance is not absent but appears to be abruptly shifting form.

The fears of humans don't always coincide with those of leatherbacks, though. Their nesting cycles are determined by the availability of prey. Only after meeting their own needs do they start storing energy for the intensive work of producing eggs and traveling to nesting beaches. The time between visits is thus different in different places. While eastern Pacifics take an average of three to four years to return, the Atlantics take an average of only two years (Stewart et al. 2010:272). These different rhythms are thought to reflect the varying levels of unpredictability each face. The "more consistent foraging environment in the Atlantic basin," and thus the shorter time between nestings, may be one reason why the population there has a more positive outlook for recovery (Stewart et al. 2010:272). Increased jellyfish blooms may remake these cycles and transform the rhythms of leatherbacks' lives. Being able to build their energy reserves more quickly could allow more frequent returns to nesting beaches and larger clutches (Stewart et al. 2010:272). Oceans filled with hauntings may replenish themselves, even yet.

Still, it is hard for conservation researchers to know. Data on jellyfish is patchy and often anecdotal. Their eerie physicality—so incorporeal that they are shredded by sampling nets, so massive that they can capsize research boats—combines with their unpredictable and polymorphous life cycles to discourage researchers from taking them up as objects of study (Schrope 2012). Lacking the time, money, methods, and inclination, Western science has shied away from learning what makes jellyfish tick.

Putting off the task of addressing the difficult questions that animals pose is not unique to conservation (Buchanan 2007), but not making the time threatens to break time. In both scientific articles (for example, Richardson et al. 2009) and the popular press (for example, Gershwin 2013), the rise of jellyfish threatens to unmake time's supposed dependability and calculability. The fear is that jellyfish may become so dominant that a regime shift could replace fish with jellyfish as "an alternative stable state in marine ecosystems" (Richardson et al. 2009:313). Relinquishing its implacable forward movement, time (whose time?) threatens to stall and begin to run in reverse, looping the Anthropocene back around into the Cambrian (Richardson et al. 2009:317).[12] But it's hard to tell. Jellyfish are not included in the models, and simulations can't be run (Richardson et al. 2009:320).

Unaware of human imaginings, jellyfish bodies react to the cascades of transformations altering the seas. They are not bound to our clacking bones, with their repeated incantation that everything will continue as it has ever done. Instead, they have heard the perfect harmony sung by intertwining rhythms—overfishing, eutrophication, climate change, translocation of invasive species, and seabed destruction (Purcell 2012). They respond, move, bloom, die, and wait—already reflecting back the times before anyone knew to look.

*   *   *

How long does it take to learn how to tell time differently? To evolve the sharp senses that are able to tune into multiple, contradictory rhythms, here, now, in our time of extinctions?

A quick glance at a clock face does not suffice. Jaguars learn to tell time with turtles over years and generations. Can we even imagine how long it took leatherbacks to tell time with jellyfish? We'll probably never know; these processes are shrouded in deep time and only occasionally read through inscriptions on rocks. We are, however, able to witness a new relationship forming knots in the time of leatherbacks, a geological moment happening right before our eyes.

In 1968, an autopsy conducted on a leatherback gives the first recorded instance of plastics being found in the animal's gastrointestinal tract (Mrosovsky, Ryan, and James 2009:288), offering a tentative date for the beginning of their fraught relationship. Since then, just over 35 percent of leatherback autopsies have revealed plastics in their guts, and, of these, they were the likely cause of death in around 9 percent of cases (Mrosovsky, Ryan, and James 2009:288). Plastic may kill only a few outright, but this new relationship adds another indeterminate cadence to the lives of leatherbacks.

Trapped in the turtles' intestines, plastics slow the absorption of nutrients. The hope for increased nestings as a result of increased jellyfish populations is now tempered by an opposing rhythm. Abundance of a food source is no help if the ability to digest it is reduced (Mrosovsky, Ryan, and James 2009:288). Here, then, is a new impediment that must somehow be coordinated with. Yet another fraught and fragile shared moment being created in a time of extinctions. How far into the future it will extend can't yet be said. As Alan Weisman (2007) writes, "Plastics haven't been around long enough yet for us to know how long they are going to be around for."

Ask why leatherbacks eat plastic, and the obvious response seems to be that the floating, bilious plastic bags have simply been mistaken for jellyfish. But ask *when* leatherbacks eat plastic, and the story becomes more complicated and more interesting. One suggestion from research done in the Gulf of Gascony is that as their jellyfish prey decreases, leatherbacks' intake of plastic increases (Duguy, Morinière, and Meunier 2000). In an abundance of jellyfish, there is not much reason to risk trying this strange new variety of prey. But hunger shifts time, and once steady, predictable relationships give way to uncertain futures.

Continued life depends on risk taking, on changing and adapting. The jaguar knows this, and so do leatherbacks. Would leatherbacks be here today if their own ancestors hadn't taken a risk and found ways of forging a beneficial relationship with toxin-laden jellyfish? By doing so, they were able to gift to their descendants a niche coveted by few other creatures (Mrosovsky, Ryan, and James 2009:287). Faced with its own new and unusual prey, the leatherback's body is again being pushed to find novel solutions. And so, hungry and more open to forging new relationships, the leatherback takes a chance and bites.

MONDAY, MARCH 10, 2014

9:20 A.M.

I am in Edinburgh, trying to write about leatherbacks again, but for the second time in a week the flesh of my palm is burning. This morning, on my way into work, a van came so close to my bike that I only had to reach out slightly to hit it in warning. I reacted so quickly that there was no time for thought, only feeling—threat, fragility, anger, self-righteousness. Knowledge of my right to be on the road turned visceral, demanding space and demanding respect.

While the taxi that I lashed out at a couple of days ago moved aside, today the driver and his passenger only looked at me blankly. Rather than delay their journey slightly, they were intent on getting through the space I was taking up and being on their way. They moved even closer, and I fell back, a slower traveler's demand for space and time overwhelmed by the demands of others.

Delay weaves its way through much of the research on leatherback conservation. The example of the torturous passage of U.S. legislation on turtle-excluder devices, which reduce the number of turtles drowning in fishing nets, is one I've written about before (Bastian 2012:44–45). General admonishments to avoid these untimely uses of time, and to work quickly and efficiently, seem to forget that these positives also cast the shadow of their negative image. After all, the delay for the turtles was justified by shrimpers' own seeming efficiencies. And today in the traffic, the focal point provided by the conjunction of destination, traffic movements, and desired arrival time obscures everything else. Time narrows, and the expansive flow that might accommodate others is funneled away by the rush of battling through all that hinders you.

Take the risk, I tell myself; follow your own time; do it differently somehow. And so, trying to avoid the focus that loses perspective, I start out each day with a reminder to go at my own pace. It becomes a mantra, "Go at your own pace; go at your own pace." But still I feel everyone's time pressing in on me. It starts to become me, and suddenly I'm chasing my own deadlines, arbitrary though they are, and the living beings around me become obstacles instead of fellow travelers. Pedestrians scurry across the road in front of me, knowing better than I do that they won't be given time to inhabit this space with others. Try as I might, I lose the expansiveness I promised I'd hold onto, and my burning palm reminds me just how far it slipped away. My time is not my own; it is given to me, absorbed by me, and offered back to the world through me.

Sitting here now at work I'm distracted, and my hand hurts, so I'm flicking around the Internet, trying to find a way of summoning up a connection. YouTube offers me the perfect link bait—leatherback rescues. Quite amenably, I bite and am reeled in.

Jumping from Newfoundland to Florida to Grenada, I watch people scrambling to help the tangled and the stranded. Fear and concern lapping against each other as they try to figure out how to return this large strange creature safely back to the oceans (for example, alinapphotography 2012; Vincent 2013). Groups of passersby collect around the scene, plans and destinations forgotten as the drama unfolds. Rusty knives, tarps, and ropes are pressed into action, and eventually the turtle is freed. Kind shouts follow—"Get going buddy"—and, not quite ready to end the moment of connection and concern, those filming continue to scan the water hoping to see it safely on its way. Eventually, in boats and on beaches, those who stopped to help are released back into their own lives and times.

Turning back to my pile of articles, I read of another video, although in this one the turtle is an obstacle to time, rather than the opening to a new one. Randell Arauz has been collecting stories of leatherbacks along the Pacific coast of South America. His report lists the number of leatherbacks killed by longlines and gill nets, and records attitudes toward interactions with leatherbacks, seeking to understand when a turtle is saved and when it isn't. He mentions a film that shows a fisherman dealing with a leatherback caught in the lines. The fisherman raises his machete to cut off a flipper so he can retrieve the hooks "in an easier and faster fashion, before being stopped" (Arauz 1999:25). Given that many turtles captured by longlines may be found alive, Arauz sees in this moment the possibility for learning to tell time differently. Careful attention could reduce the number of turtles who die from the injuries sustained during gear removal (Arauz 1999:25). Thus while Arauz (1999) suggests that many are responsible for stopping the decline of turtle numbers he writes that it is fishers "who will have the ultimate responsibility during fishing operations at high-seas, of saving that turtle on the hook" (26)!

While Shillinger and his colleagues (2011) hope for a discon-nection, Arauz (1999) invests in the moment of connection as the time when conservation might do some of its most crucial work. The steady tick of predictability and calculability that echoed through the planned turtle watch becomes a background note. Instead, Arauz turns toward the same tick that chivvied me into action early on a Sydney morning. Under a watchful gaze, those who are out of sync are insistently reminded to adjust, catch up, keep to time. So many of us then chase the lie that all that is needed for proper coordination is for individuals to appropri-ately calibrate themselves with the correct forms of time (Sharma 2014:138).

But can taking the time to recalibrate to a time of care be done alone? As Maria Puig de la Bellacasa (2012:198) suggests, acts of care are embedded in interdependent worlds, and those expected to care may often be laboring under conditions of exploitation and domination. The tourist on the beach wedging the tarp under the stranded turtle and the fisher out at sea are enmeshed in very dif-ferent webs of time, with different rhythms, expectations, futures, and pasts, pressing in on each of them in different ways. Adjusting to a time in which fishers can be "patient enough to release hooked turtles, untangle them, or use techniques to safely release hooked turtles" (Arauz 1999:25) may involve more tangles than just those accessible to the fishers alone. As Sarah Sharma (2014) argues, time is not "singularly yours or mine for the taking but [is] uncompromisingly tethered and collective" (150).

A jaguar's time is tethered to its shifting prey; a turtle's, to the amount of plastic in its gut, just two threads among many. These stories suggest that learning to tell time differently is both a collective risk and a collective task, though not in the same way for

everyone. After all, it's easy to focus on the single-minded fisherman wielding his machete, but this tracing of connections with
leatherbacks will also bring me face to face with the narrowings of
time fostered by my own trade.

A fisherman is complaining about the lack of response from researchers. He has returned at least thirty tags to a research project
in Costa Rica and has never had a reply. Arauz (1999:21) delicately
describes the fisherman's reaction as "discouragement" over this
lack of interest. Originally from Costa Rica, where he participated
in an environmental education program, he is now involved in
longline fishing in Ecuador. He has taken this education to heart
and tries to take care of any turtles he encounters. But his efforts
to help support the continuation of shared futures between turtles
and humans are met with a foggy uncertainty.

In my now unsteady pile of research papers and reports, I follow this thread all the way to Canada, where fishers there, too, have
received no feedback on tags and no follow-up after "spending
hours hauling a full-size whale to shore" for researchers to study
(Martin and James 2005:114). Conservationists trying to do things
differently find that employers and funders are insensible to the
multiple, contradictory rhythms involved in building ongoing
communities of concern. They face the continuing challenge of
"convincing funding agencies that are conditioned to support
traditional research that funding 'softer' aspects of a conservation programme, like community outreach, is supporting science"
(Martin and James 2005:113; see also Delgado and Nichols
2005:96). Cutting time back to its bones may seem to support
staying consistently on target, but it leaves the remnants of the
careful responses of others trailing in its wake.

Not everywhere, though. Other threads of time belie the clock's
claim that one time can encompass all. Kathleen Martin and her
colleagues (Martin and James 2005) are involved in the Nova

Scotia Leatherback Turtle Working Group (NSLTWG), which works closely with local fishers on conservation projects. Many are no longer able to hunt for swordfish, which have become increasingly rare, and so the years of cultivating particular embodiments are turned to other uses. Forging new futures, they now go "turtling," working with conservation scientists to learn more about the behavior of leatherbacks in Canadian waters. As Martin and James (2005) write, "The ability to spot leatherback turtles at sea requires observational abilities that only those who have fished on the ocean for years can cultivate" (113). Indeed, like the jellyfish of the northeastern Atlantic, until the fishers of the NSLTWG turned their swordfish-trained eyes to turtle spotting, the presence of leatherbacks in those waters had never been scientifically proven.

By working closely with local volunteers and seeking to build trust among communities whose interests are not always aligned, Martin and her colleagues make time for careful relationship. But this time carries consequences—academic productivity, status, peer recognition are all put at risk (see also Campbell 2005). For the fishers, however, breaking professional codes by being involved in voluntary conservation work is to risk suspicion, social exclusion, even death (Delgado and Nichols 2005:99). For both sets of partners, taking time involves falling out of the complex, but also enfolding, rhythms that bind communities together. But the same risk is not shared by everyone, and the greater risk cannot always be paid back or balanced out. As Martin and James (2005) write, in relation to the fishers they work with, "There is no way to 'repay' the cultural risk entailed in this kind of action" (115n.1).

Discussing the violence entwined with care in conservation, van Dooren (2014a) writes that it is always important to ask, "What am I really caring for, why, and at what cost to whom?" Likewise, Sharma (2014) reminds us to ask, "What new forms of vulnerability are necessitated by the production of temporal novelties" (150)? What were those bones I worried over, sitting on the carpet in a faraway Sydney? How many other lives were entangled with them while I sat there, intoxicated by the way they seemed to hail me alone?

MONDAY, AUGUST 4, 2014

4:21 P.M.

The writing that started with such anxiety, after being put off for too long, is nearing completion. Layers of deadlines for conferences and seminars, drafts and redrafts, comments and criticisms have worked it all into a kind of coherence. Throughout it all, leatherbacks have surfaced in unexpected places, opening up shared worlds in which the calculability of time is disrupted, its seemingly implacable forward movement turned on its head and admonishments to work faster, be more consistent, and be more focused are not able to provide the time needed to solve the problems at hand. Rather than connecting with "present temporalities, localities, and relationalities," the time given by leatherbacks has rendered each of them unfamiliar.

Sifting through news items reporting on others' encounters traces a similar sense of estrangement. Stories of sightings, rescues, and nestings—all accompanied by astonishment that such a creature should appear *here*. There were the "completely baffled" experts trying to work out how a dog walker could find fresh leatherback eggs on a beach on Jersey, one of the Channel Islands (BBC 2013). And wildlife watchers off the coast of Cornwall talking about the "enormous privilege" of seeing one so close to land (Lester 2013). A turtle has even been sighted hauling up on the beach in England's Blackpool (Cooke 2010). Closer to my old home in Australia, so far removed (or so I thought) from leatherback haunts, a dead turtle, probably killed by a boat strike, had drifted ashore near Byron Bay, and it was "believed to be the first time in 17 years this breed of turtle has been seen on the East Coast" (Kinninment 2013). Another was seen alive in Melbourne's Port Philip Bay (Florance 2014). Sharing my confusion were reporters in Balatan in the Philippines who wondered why an animal "only seen in the Atlantic waters in Europe" would be found tangled in local fishing gear (Sales 2013).

To encounter a leatherback, then, might actually mean having one's sense of place and time disoriented. As Martin attests, "You really feel like you're being blessed by the primeval, you know, this is an animal who has been around for 150 million years—since the T. rex was on Earth, leatherbacks have been with us—it's such a privilege to see that and have that sense of being tied into a world that is so much older than you are, and so much bigger, and just more mysterious" (quoted in CBC 2014). Envoys from the last great extinction event, a leatherback encounter may offer a moment that bones cannot touch, a moment that squawks and shuffles and captivates.

But my ticking clock won't give up easily. It's now 11:28 P.M., and I'm on the brink of falling back into the untimeliness that started all of this. There are so many tangles, knots, and threads that I'm not sure which ones I should track down, tidy up, or cut away.

I need some fresh air. So I quietly unlock the front door and step outside. The street lights give everything an orange glow, and I can hear faint sounds of traffic on the roads. The Water of Leith is close by, and I start to follow it along as it runs through Edinburgh's suburbs. Along and along in the cool darkness. When I get to Inverleith Park, I leave the river and follow the roads straight down to Granton Harbour, and here I stop, looking out over the water.

I look for them; and, don't see any yet. But I might.

Leatherbacks have been here recently.[13]

While I wait I pull out the clock's bones from my pockets.

It's time to let them go,

so I lay them carefully on the surface of the water.

For a moment they just float there,

but, after a little while,

they start to grow into each other, stretching flesh and sprouting wings

before heaving up out of the water and soaring lazily out to sea.

NOTES

1. Although see also Kevin Birth's (2014) critique of the way the metaphor of the clock has led to misunderstandings of how these body "clocks" work.

2. See, for example, specifically "Weekday Calculator—What Day Is This Date?" http://www.timeand date.com/date/weekday.html; and http://www.moonpage.com/index.html.

3. Research also suggests that the hatchlings do their own forms of synchronizing, calling to one another while still within their shells in order to coordinate their crawl to the ocean (Ferrara et al. 2014).

4. For information, see NOAA Fisheries, "TurtleWatch," National Oceanic and Atmospheric Administration, http://www.pifsc.noaa.gov /eod/turtlewatch.php.

5. Since I wrote this essay, the Hawaiian TurtleWatch mapping tool has been extended to cover leatherback interactions as well (Howell et al. 2015).

6. PTT stands for Platform Transmitter Terminals, which are used with the Argos tracking and monitoring system. For a discussion of the system's development, see Benson (2012).

7. For information, see "Intra-action: Multispecies Becomings in the Anthropocene" [exhibition at the conference of the Australian Animal Studies Group, University of Sydney, July 8–10, 2013], http://intraac tionart.com/.

8. This metaphor is inspired by Deborah Bird Rose's (2012) interest in "add[ing] flesh to the relatively abstracted analysis of kinds of time and patterns that connect" (128).

9. I'm thinking here of Birth's (2014) use of the term "triangulation," where time is reckoned by "relating the intersection of different timing or cyclical phenomena," similar to the "navigational practice of locating one's position in space by reference to three or more known locations" (318).

10. For a more recent overview, see Gibbons and Richardson (2013).

11. More recent literature questions this, suggesting that while there have been increases in localized blooms, there is insufficient research to tell whether there are global trends toward population increase (Condon et al. 2013).

12. For a critique of the use of these sorts of temporal moves in scientific research, see Schrader (2012).

13. For a map of the sightings of leatherbacks around the British Isles, see "Grid Map for Dermochelys coriacea (Vandelli, 1761) [Leathery Turtle]," NBN Gateway, https://data.nbn.org.uk/Taxa/NBNSYS 0000188646/Grid_Map.

REFERENCES

Adam, Barbara. 1998. *Timescapes of Modernity: The Environment and Invisible Hazards*. New York: Routledge.

alinapphotography. 2012. "Leatherback Mama Turtle Rescue in Highland Beach, FL" [part 2 of 7]. YouTube. https://www.youtube .com/watch?v=0YfK2Th-_VY.

Arauz, Randall. 1999. "Description of the Eastern Pacific High-Seas Longline and Coastal Gillnet Swordfish Fisheries of South America, Including Sea Turtle Interactions, and Management Recommendations." Report submitted to James R. Spotila, Drexel University. Sea Turtle Restoration Project, Turtle Island Restoration Network.

Arroyo-Arce, Stephanny, James Guilder, and Roberto Salom-Pérez. 2014. "Habitat Features Influencing Jaguar *Panthera onca* (Carnivora: Felidae) Occupancy in Tortuguero National Park, Costa Rica." *International Journal of Tropical Biology and Conservation* 62, no. 4:1449–1458.

Bailey, Helen, Scott R. Benson, George L. Shillinger, Steven J. Bograd, Peter H. Dutton, Scott A. Eckert, Stephen J. Morreale, Frank V. Paladino, Tomoharu Eguchi, David G. Foley, Barbara A. Block, Rotney Piedra, Creusa Hitipeuw, Ricardo F. Tapilatu, and James R. Spotila. 2012. "Identification of Distinct Movement Patterns in Pacific Leatherback Turtle Populations Influenced by Ocean Conditions." *Ecological Applications* 22, no. 3:735–747. doi:10.1890/11-0633.

Bailey, Helen, Sabrina Fossette, Steven J. Bograd, George L. Shillinger, Alan M. Swithenbank, Jean-Yves Georges, Philippe Gaspar, K. H. Patrik Strömberg, Frank V. Paladino, James R. Spotila, Barbara A. Block, and Graeme C. Hays. 2012. "Movement Patterns for a Critically, Endangered Species, the Leatherback Turtle (*Dermochelys coriacea*), Linked to Foraging Success and Population Status." *PLoS ONE* 7, no. 5:e36401.

Bastian, Michelle. 2012. "Fatally Confused: Telling the Time in the Midst of Ecological Crises." *Environmental Philosophy* 9, no. 1:23–48.

——. 2014. "Time." In *Migration: The COMPAS Anthology*, edited by Bridget Anderson and Michael Keith, 62. Oxford: COMPAS.

——. Forthcoming. "Liberating Clocks: Developing a Critical Horology to Rethink the Potential of Clock Time." *new formations: a journal of culture / theory / politics*.

BBC. 2013. "Jersey Resident 'Finds Rare Leatherback Turtle Egg on Beach.'" BBC News, June 11. http://www.bbc.co.uk/news/world -europe-jersey-22853538.

Benson, Etienne. 2012. "One Infrastructure, Many Global Visions: The Commercialization and Diversification of Argos, a Satellite-Based Environmental Surveillance System." *Social Studies of Science* 42, no. 6:843–868. doi:10.1177/0306312712457851.

Benson, Scott R., Tomoharu Eguchi, Dave G. Foley, Karin A. Forney, Helen Bailey, Creusa Hitipeuw, Betuel P. Samber, Ricardo F. Tapilatu, Vagi Rei, Peter Ramohia, John Pita, and Peter H. Dutton. 2011. "Large-Scale Movements and High-Use Areas of Western Pacific Leatherback Turtles, *Dermochelys coriacea.*" *Ecosphere* 2, no. 7:art84. doi:10.1890/ES11-00053.1.

Birth, Kevin K. 2014. "Non-Clocklike Features of Psychological Timing and Alternatives to the Clock Metaphor." *Timing & Time Perception* 2, no. 3:312–324. doi:10.1163/22134468-00002029.

Bjorndal, Karen A., and Alan B. Bolten. 2003. "From Ghosts to Key Species: Restoring Sea Turtle Populations to Fulfill Their Ecological Roles." *Marine Turtle Newsletter* 100:16–21.

Buchanan, Brett. 2007. "The Time of the Animal." *PhaenEx* 2, no. 2: 61–80.

Campbell, Lisa M. 2005. "Overcoming Obstacles to Interdisciplinary Research." *Conservation Biology* 19, no. 2:574–577. doi:10.1111/j.1523-1739 .2005.00058.x.

——. 2007. "Local Conservation Practice and Global Discourse: A Political Ecology of Sea Turtle Conservation." *Annals of the Association of American Geographers* 97, no. 2:313–334.

Campbell, Lisa M., and Christina Smith. 2005. "Volunteering for Sea Turtles? Characteristics and Motives of Volunteers Working with the Caribbean Conservation Corporation in Tortuguero, Costa Rica." In "Marine Turtles as Flagship," edited by Jack Frazier. Special issue, *MAST: Maritime Studies* 4, no. 1:169–193.

Carrillo, Eduardo, Todd K. Fuller, and Joel C. Saenz. 2009. "Jaguar (*Panthera onca*) Hunting Activity: Effects of Prey Distribution and Availability." *Journal of Tropical Ecology* 25, no. 5:563–567. doi:10.1017 /S0266467409990137.

Carrillo, Eduardo, Joel C. Saenz, and Todd K. Fuller. 2002. "Movements and Activities of White-Lipped Peccaries in Corcovado National Park, Costa Rica." *Biological Conservation* 108, no. 3:317–324. doi:http://dx.doi.org/10.1016/S0006-3207(02)00118-0.

CBC. 2014. "Endangered Leatherback Turtle Speared by Swordfish Survives." CBC News, June 19. http://www.cbc.ca/news/canada /nova-scotia/endangered-leatherback-turtle-speared-by-swordfish -survives-1.2680768.

Condon, Robert H., Carlos M. Duarte, Kylie A. Pitt, Kelly L. Robinson, Cathy H. Lucas, Kelly R. Sutherland, Hermes W. Mianzan, Molly Bogeberg, Jennifer E. Purcell, Mary Beth Decker, Shin-ichi Uye, Laurence P. Madin, Richard D. Brodeur, Steven H. D. Haddock, Alenka Malej, Gregory D. Parry, Elena Eriksen, Javier Quiñones, Marcelo Acha, Michel Harvey, James M. Arthur, and William M. Graham. 2013. "Recurrent Jellyfish Blooms Are a Consequence of Global Oscillations." *Proceedings of the National Academy of Sciences* 110, no. 3:1000–1005. doi:10.1073/pnas.1210920110.

Cooke, Jeremy. 2010. "Giant Leatherback Turtle Visits Beach near Blackpool." BBC News, June 10. http://www.bbc.co.uk/news /10289590.

Danigelis, Alyssa. 2013. "Jellyfish Shut Down Swedish Nuclear Reactor." *Seeker*, October 2. http://www.seeker.com/jellyfish-shut-down -swedish-nuclear-reactor-1767888308.html.

Delgado, Stephen, and Wallace J. Nichols. 2005. "Saving Sea Turtles from the Ground Up: Awakening Sea Turtle Conservation in North-Western Mexico." In "Marine Turtles as Flagship," edited by Jack Frazier. Special issue, *MAST: Maritime Studies* 4, no. 1:89–104.

Doyle, Thomas K., Henk De Haas, Don Cotton, Boris Dorschel, Valerie Cummins, Jonathan D. R. Houghton, John Davenport, and Graeme C. Hays. 2008. "Widespread Occurrence of the Jellyfish *Pelagia noctiluca* in Irish Coastal and Shelf Waters." *Journal of Plankton Research* 30, no. 8:963–968. doi:10.1093/plankt/fbn052.

Duguy, R., P. Morinière, and A. Meunier. 2000. "L'ingestion des déchets flottants par la tortue luth *Dermochelys coriacea* (Vandelli, 1761) dans le golfe de Gascogne." *Annales de la Société des Sciences Naturelles de la Charente-Maritimes* 8, no. 9:1035–1038.

EFE. 2014. "Number of Leatherback Turtle Nests on Puerto Rico Beaches Soars." Fox News Latino, June 21. http://latino.foxnews. com/latino/news/2014/06/21/number-leatherback-turtle-nests-on-puerto-rico-beaches-soars/.

Ferrara, Camila R., Richard C. Vogt, Martha R. Harfush, Renata S. Sousa-Lima, Ernesto Albavera, and Alejandro Tavera. 2014. "First Evidence of Leatherback Turtle (*Dermochelys coriacea*) Embryos and Hatchlings Emitting Sounds." *Chelonian Conservation and Biology* 13, no. 1:110–114. doi:10.2744/ccb-1045.1.

Florance, Loretta. 2014. "Endangered Leatherback Sea Turtle Spotted in Melbourne's Port Phillip Bay." ABC News, December 21. http://www.abc.net.au/news/2014-12-20/endangered-leatherback-turtle-spotted-in-port-phillip-bay/5981088.

Gershwin, Lisa-Ann. 2013. *Stung! On Jellyfish Blooms and the Future of the Ocean.* Chicago: University of Chicago Press.

Gibbons, Mark J., and Anthony J. Richardson. 2013. "Beyond the Jellyfish Joyride and Global Oscillations: Advancing Jellyfish Research." *Journal of Plankton Research* 35, no. 5:929–938. doi:10.1093/plankt/fbt063.

Gordon, Lucía Galeán, and Emma Harrison. 2012. *Report on the 2011 Leatherback Program at Tortuguero, Costa Rica.* San Pedro, Costa Rica: Sea Turtle Conservancy.

Guilder, James, Benjamin Barca, Stephanny Arroyo-Arce, Roberto Gramajo, and Roberto Salom-Pérez. 2015. "Jaguars (*Panthera onca*) Increase Kill Utilization Rates and Share Prey in Response to Seasonal Fluctuations in Nesting Green Turtle (*Chelonia mydas mydas*) Abundance in Tortuguero National Park, Costa Rica." *Mammalian Biology–Zeitschrift für Säugetierkunde* 80, no.2:65–72. doi:http://dx.doi.org/10.1016/j.mambio.2014.11.005.

Hay, Steve. 2006. "Marine Ecology: Gelatinous Bells May Ring Change in Marine Ecosystems." *Current Biology* 16, no. 17:R679-R682. doi:http://dx.doi.org/10.1016/j.cub.2006.08.010.

Hays, Graeme C. 2008. "Sea Turtles: A Review of Some Key Recent Discoveries and Remaining Questions." *Journal of Experimental Marine Biology and Ecology* 356, nos. 1–2:1–7. doi:http://dx.doi.org/10.1016/j.jembe.2007.12.016.

Heaslip, Susan G., Sara J. Iverson, W. Don Bowen, and Michael C. James. 2012. "Jellyfish Support High Energy Intake of Leatherback Sea Turtles (*Dermochelys coriacea*): Video Evidence from Animal-Borne Cameras." *PLoS ONE* 7, no. 3:e33259. doi:10.1371/journal.pone.0033259.

Houghton, Jonathan D. R., Thomas K. Doyle, Mark W. Wilson, John Davenport, and Graeme C. Hays. 2006. "Jellyfish Aggregations and Leatherback Turtle Foraging Patterns in a Temperate Coastal Environment." *Ecology* 87, no. 8:1967–1972. doi:10.1890/0012-9658 (2006)87[1967:jaaltf]2.0.co;2.

Howell, Evan A., Aimee Hoover, Scott R. Benson, Helen Bailey, Jeffrey J. Polovina, Jeffrey A. Seminoff, and Peter H. Dutton. 2015. "Enhancing the TurtleWatch Product for Leatherback Sea Turtles, a Dynamic Habitat Model for Ecosystem-Based Management." *Fisheries Oceanography* 24, no. 1:1–12. doi:10.1111/fog.12092.

James, Michael C., C. Andrea Ottensmeyer, A. Ransom, and Andrew E. Myers. 2005. "Identification of High-Use Habitat and Threats to Leatherback Sea Turtles in Northern Waters: New Directions for Conservation." *Ecology Letters* 8:195–201. doi:10.1111 /j.1461-0248.2004.00710.x.

Kinninment, Megan. 2013. "Rare Giant Turtle Washes Up at Suffolk Park." *Northern Star* (Lismore, Australia), November 24. http://www.northernstar.com.au/news/a-rare-leatherback-turtle-has-been-found-washed-up/2093769/.

Kwek, Glenda. 2011. "Jellyfish Force Shutdown of Power Plants." *Sydney Morning Herald*, July 11. http://www.smh.com.au/environment /jellyfish-force-shutdown-of-power-plants-20110711-1haa6.html.

Lester, Nick. 2013. "Are You a Little Lost? Tourists Picture Giant Leatherback Turtle 3,000 Miles from Home as It Makes Rare

Appearance in British Waters in the Hunt for Jellyfish." *Daily Mail*, September 13. http://www.dailymail.co.uk/news/article-2419937 /Giant-leatherback-turtle-spotted-makes-rare-appearance-British-waters-hunt-jellyfish.html.

Mann, Adam. 2014. "How the U.S. Built the World's Most Ridiculously Accurate Atomic Clock." *Wired*, April 4. http://www.wired. com/2014/04/nist-atomic-clock/.

Martin, Kathleen, and Michael James. 2005. "The Need for Altruism: Engendering a Stewardship Ethic Amongst Fishers for the Conservation of Sea Turtles in Canada." In "Marine Turtles as Flagship," edited by Jack Frazier. Special issue, *MAST: Maritime Studies* 4, no. 1:105–118.

Mrosovsky, N., Geraldine D. Ryan, and Michael C. James. 2009. "Leatherback Turtles: The Menace of Plastic." *Marine Pollution Bulletin* 58, no. 2:287–289.

Pauly, Daniel. 1995. "Anecdotes and the Shifting Baseline Syndrome of Fisheries." *Trends in Ecology & Evolution* 10, no. 10:430. doi:10.1016 /s0169-5347(00)89171-5.

Pittendrigh, Colin S. 1981. "Circadian Systems: Entrainment." In *Biological Rhythms*, edited by Jürgen Aschoff, 95–124. New York: Springer.

Puig de la Bellacasa, María. 2012. "'Nothing comes without its world': Thinking with Care." *Sociological Review* 60, no. 2:197–216. doi:10.1111 /j.1467-954X.2012.02070.x.

Purcell, Jennifer E. 2012. "Jellyfish and Ctenophore Blooms Coincide with Human Proliferations and Environmental Perturbations." *Annual Review of Marine Science* 4, no.1:209–235. doi:10.1146/annurev -marine-120709-142751.

Purcell, Jennifer E., Shin-ichi Uye, and Wen-Tseng Lo. 2007. "Anthropogenic Causes of Jellyfish Blooms and Their Direct Consequences for Humans: A Review." *Marine Ecology Progress Series* 350:153–174.

Richardson, Anthony J., Andrew Bakun, Graeme C. Hays, and Mark J. Gibbons. 2009. "The Jellyfish Joyride: Causes, Consequences and

Management Responses to a More Gelatinous Future." *Trends in Ecology & Evolution* 24, no. 6:312–322.

Roberts, Callum. 2007. *The Unnatural History of the Sea: The Past and Future of Humanity and Fishing*. London: Gaia.

Rose, Deborah Bird. 2012. "Multispecies Knots of Ethical Time." *Environmental Philosophy* 9, no. 1:127–140.

——. 2013. "Slowly—Writing into the Anthropocene." In "Writing Creates Ecology and Ecology Creates Writing," edited by Martin Harrison, Deborah Bird Rose, Lorraine Shannon, and Kim Satchell. Special issue, *TEXT* 20:1–14.

Saba, Vincent S., George L. Shillinger, Alan M. Swithenbank, Barbara A. Block, James R. Spotila, John A. Musick, and Frank V. Paladino. 2008. "An Oceanographic Context for the Foraging Ecology of Eastern Pacific Leatherback Turtles: Consequences of ENSO." *Deep Sea Research Part I: Oceanographic Research Papers* 55, no. 5:646–660.

Sales, Sonny. 2013. "Rare Leatherback Turtle Released to Sea." *Vox Bikol* (Naga City, Philippines), October 2. http://www.voxbikol.com/article/rare-leatherback-turtle-released-sea.

Schrader, Astrid. 2012. "The Time of Slime: Anthropocentrism in Harmful Algal Research." *Environmental Philosophy* 9, no. 1:71–93.

Schrope, Mark. 2012. "Marine Ecology: Attack of the Blobs." *Nature* 482:20–21. doi:10.1038/482020a.

Sharma, Sarah. 2014. *In the Meantime: Temporality and Cultural Politics*. Durham, N.C.: Duke University Press.

Shillinger, George L., Emanuele Di Lorenzo, Hao Luo, Steven J. Bograd, Elliott L. Hazen, Helen Bailey, and James R. Spotila. 2012. "On the Dispersal of Leatherback Turtle Hatchlings from Mesoamerican Nesting Beaches." *Proceedings of the Royal Society B: Biological Sciences* 279, no. 1737:2391–2395. doi:10.1098/rspb.2011.2348.

Shillinger, George L., Daniel M. Palacios, Helen Bailey, Steven J. Bograd, Alan M. Swithenbank, Philippe Gaspar, Bryan P. Wallace, James R. Spotila, Frank V. Paladino, Rotney Piedra, Scott A. Eckert, and

Barbara A. Block. 2008. "Persistent Leatherback Turtle Migrations Present Opportunities for Conservation." *PLoS Biology* 6, no. 7:e171.

Shillinger, George L., Alan M. Swithenbank, Helen Bailey, Steven J. Bograd, Michael R. Castelton, Bryan P. Wallace, James R. Spotila, Frank V. Paladino, Rotney Piedra, and Barbara A. Block. 2011. "Vertical and Horizontal Habitat Preferences of Post-Nesting Leatherback Turtles in the South Pacific Ocean." *Marine Ecology Progress Series* 422:275–289. doi:10.3354/meps08884.

Shillinger, George L., Alan M. Swithenbank, Steven J. Bograd, Helen Bailey, Michael R. Castelton, Bryan P. Wallace, James R. Spotila, Frank V. Paladino, Rotney Piedra, and Barbara A. Block. 2010. "Identification of High-Use Interesting Habitats for Eastern Pacific Leatherback Turtles: Role of the Environment and Implications for Conservation." *Endangered Species Research* 10:215–232. doi:10.3354/esr00251.

Spotila, James R., Richard D. Reina, Anthony C. Steyermark, Pamela T. Plotkin, and Frank V. Paladino. 2000. "Pacific Leatherback Turtles Face Extinction." *Nature* 405:529–530. doi:10.1038/35014729.

Stanford University. 2008. "Leatherback Turtles' Newly Discovered Migration Route May Be Roadmap to Salvation." Press release, July 15. http://news.stanford.edu/pr/2008/pr-leatherb-072308.html.

Stewart, Kelly, Michelle Sims, Anne Meylan, Blair Witherington, Beth Brost, and Larry B. Crowder. 2010. "Leatherback Nests Increasing Significantly in Florida, USA: Trends Assessed over 30 Years Using Multilevel Modeling." *Ecological Applications* 21, no. 1:263–273. doi:10.1890/09-1838.1.

Taylor, Hollis. 2008. "Decoding the Song of the Pied Butcherbird: An Initial Survey." *Transcultural Music Review* 12, no. 2:1–30.

——. 2013. "Connecting Interdisciplinary Dots: Songbirds, 'White Rats' and Human Exceptionalism." *Social Science Information* 52, no. 2: 287–306. doi:10.1177/0539018413477520.

van Dooren, Thom. 2014a. "'Care' in the Living Lexicon for the Environmental Humanities." *Environmental Humanities* 5:291–294

——. 2014b. *Flight Ways: Life and Loss at the Edge of Extinction.* New York: Columbia University Press.

Veríssimo, D., D. A. Jones, R. Chaverri, and S. R. Meyer. 2012. "Jaguar *Panthera onca* Predation of Marine Turtles: Conflict Between Flagship Species in Tortuguero, Costa Rica." *Oryx* 46, no. 3:340–347.

Vincent, Blair. 2013. "Giant Leatherback Sea Turtle Freed from Entanglement." YouTube. https://www.youtube.com/watch?v=j9CwJFCUUcI.

Weisman, Alan. 2007. "Polymers Are Forever." *Orion Magazine*, May–June.

Witt, Matthew J., Annette C. Broderick, David J. Johns, Corinne Martin, Rod Penrose, Marinus S. Hoogmoed, and Brendan J. Godley. 2007. "Prey Landscapes Help Identify Potential Foraging Habitats for Leatherback Turtles in the NE Atlantic." *Marine Ecology Progress Series* 337:231–243. doi:10.3354/meps337231.

Margaret Barnaby, *Sanguine Moon*, with two ʻalalā (*Corvus hawaiiensis*) on a loulu palm. (Woodblock print. © Margaret Barnaby)

6. SPECTRAL CROWS IN HAWAI'I

Conservation and the Work of Inheritance

THOM VAN DOOREN

I stood in the forest listening for crows. Listening and hoping, even though I knew that it was foolish. I had been led to this forest precisely because there were no longer crows here, because there were no longer free-living crows anywhere in Hawai'i. I knew that the last sighting of a crow had been made a decade earlier, in 2002, and that these birds were now extinct in the wild. But as I stood in the forest, I couldn't help but listen and hope.

I had read descriptions of crows in Hawaiian forests by eighteenth- and nineteenth-century naturalists and ornithologists, writing when these birds were still relatively common. George Munro (1944) saw them in 1891 and provided a passing reference to their graceful movements below the rain-forest canopy: birds "sail[ing] from tree to tree on motionless wings" (70). Standing in a forest at 7,000 feet elevation—in the heart of the region where they once lived—I imagined for a moment that I could see their feathered forms moving through the trees. I imagined what it would be like for the now eerily quiet forest, missing this and so many other species of birds, to once again be enlivened by such a charismatic presence.

And so, we begin with spectral crows, haunting a dying forest. This forest was itself in decline for a number of reasons, principally because of the presence of introduced ungulates, like pigs, that uproot and graze down any new vegetation. Where once there had been a lush understory beneath a tall canopy of trees, all that remained now were old trees with no new growth to replace them, and no understory to hold the soil together when it rained. The biologists I was traveling with called this a "museum forest"; others have called it a forest of the "living dead" (Sodikoff 2013). Either way, it too was perched perilously at the edge between life and death.

In a range of ways, this chapter is an exploration of the absence of Hawai'i's crows as well as some of the many contestations over, and consequences of, their potential return. In particular, I am interested in how we inherit and inhabit the legacies of the past to shape possible futures. These inheritances take many forms: from genetic material and the broader landscapes and ecological communities that we are born into, to the historical events and relationships that we re-tell and remember and that consequently guide our understandings of and actions in the world. In a time of ongoing extinction and colonization, a time in many ways characterized by interwoven patterns of biological and cultural loss, what does it mean to inherit *responsibly*? My contention is that in a "postnatural world"—one that refuses the dangerous illusion of wilderness—conservation must be rethought as a "work of inheritance."

The crow that is my guide into these questions is not just any crow. Known locally by their Hawaiian name—'alalā—these birds are forest and fruit specialists. Although they look very much like the more abundant species of crow and raven found widely around the world, behaviorally they are quite unique. 'Alalā do not seem to have taken to scavenging and a life beyond the forest. Instead, they ate flowers and fruit, insects, and occasionally other birds' eggs. As Polynesian and then European, Asian, and other peoples arrived, ʻalalā stayed in the forests even as these places were becoming less and less hospitable for them. Some forests were cleared, and others were degraded by introduced ungulates. Meanwhile, new avian diseases and predators like cats and mongooses moved in.

Eventually, roughly a decade ago, the last of the free-living ʻalalā died. Initially, only a handful of crows survived in captivity. As a result of years of captive breeding, however, there are now roughly 110 ʻalalā held in two facilities on the Big Island (Hawaiʻi) and Maui. This conservation project is a collaboration between the state and federal governments and the San Diego Zoo. Working together, they hope that one day soon these birds might be able to start being released back into the forests of the Big Island. Before this can happen, however, much remains to be done to prepare the way.

Ghosts and Co-Becoming at the Edge of Extinction

We don't know when it was, or where they came from, but at some point in the deep history of the Hawaiian Islands, crows appeared. As the islands in this volcanic chain rose above the sea, one by one countless plants, animals, and other species arrived by wave, wind, and wing and settled in. A diversity of life broke forth. Animals and plants adapted, coevolving with others over millions of years. Completely free of mammalian predators, for the longest time these were islands of immense avian diversity. Fossil records indicate that there was once a range of large, flightless birds in the islands (Steadman 2006). It is likely that in earlier times, many of these birds played important ecological roles as pollinators or seed dispersers for local plants.

Today, however, most of these birds are long gone. Of the 113 avian species known to have lived exclusively on these islands just prior to human arrival, almost two-thirds are now extinct. Of the 42 species that remain, 31 are federally listed under the Endangered Species Act (Leonard 2008). It is not hard to see why Hawai'i is regarded as one of the "extinction capitals" of the world.

As a result, the 'alalā is now the largest fruit-eating forest bird remaining anywhere in the islands—albeit only in captivity. With its passing from the forest it is thought that several plant and tree species—especially some of those with bigger fruits and seeds—may have lost their only remaining seed disperser. Under the rain-forest canopy, wide seed dispersal can be a vital component of species survival. As birds carry seeds away from their parent trees, they spread genetic diversity, they reduce competition, and they can even provide safer places for germination. For example, many Hawaiian plants can also grow epiphytically, safe from browsing ungulates, if their seeds are deposited in the canopy by birds.

Research conducted by Susan Moana Culliney and her colleagues (2012) suggests that the 'alalā may have been the last remaining seed disperser for at least three plants: ho'awa, halapepe, and the loulu palms. But dispersal is not just about movement. In addition, it seems that some of these seeds germinate better— or, in the case of ho'awa, germinate only—if the outer fruit has been removed, something that 'alalā once routinely did.

A long and intimate history of coevolution lies within these embodied affinities that bind together avian and botanical lives. Crows are nourished, plants are propagated, and in the process both species are, at least in part, constituted: their physical and behavioral forms, their *ways of life*, emerging out of generation after generation of co-evolutionary "intra-action" (Barad 2007).

'Alalā haunt the forest in another way here. Beyond my own active imagination, their spectral presence is *inscribed* in the forest landscape. Plants call out to ʻalalā, their fruiting and flowering bodies shaped by past attractions and associations that no longer exist. This "call" is not simply metaphorical; it is a fleshy, embodied, evolved, and continually reenacted semiosis.

As ʻalalā populations have declined over the past decades, the plants bound up in mutualistic relations with them have likely declined, too. Halapepe and loulu palms are themselves rare or endangered. In addition, Culliney (2011) notes with regard to hoʻawa that most of the trees encountered today are older and that there is now a "general lack of seedlings or saplings in the wild" (21). It is quite possible that these plants are now what biologists call "ecological anachronisms": species with traits that evolved in response to a relationship or an environmental condition that is no longer present (Barlow 2000; Janzen and Martin 1982). The extent to which the loss of ʻalalā has contributed to the decline of these plant species remains a topic for future study. It is clear, however, that the absence of a seed disperser can only make the future of these plant species that much more precarious. Here, we see that coevolution can switch over into coextinction; co-becoming into entangled patterns of dying-with.

Alongside plants and their forests, the disappearance of ʻalalā is also felt by local people. For some Native Hawaiians, ʻalalā are part of their cultural landscape: these birds hold stories and associations in the world. ʻAlalā is an *ʻaumakua*, or ancestral deity, for some people, and the plants and forests that might disappear or change significantly without their seed disperser are themselves also culturally significant in various ways (Culliney 2011). Many other locals are also drawn into this experience of loss. I interviewed biologists, artists, ranchers, hunters, and others, some of whom were lucky enough to remember—and so miss—the dramatic presence of these birds in the forest. Many of these people were trying in their own ways to reckon with the affective burden of living in a place in which crows are no longer present, a place in which (paraphrasing one biologist), we have lost the most intelligent and charismatic component of our forests.[1]

Here, crows, plants, people, and others are tangled up and at stake in one another. But it is the particularly historical character of these entanglements that I am interested in. More specifically, the way in which life is, at a fundamental level, grounded in rich patterns of *inheritance*. All of Earth's creatures are heirs to the long history of life on this planet. We are woven through with traces of the past: our own past, but also that of our forebears whose relationships and achievements we inherit in our genes, our cultural practices, our languages, and much more.

In recent decades, these entangled, biocultural processes of inheritance have become somewhat more readily intelligible from within the biological sciences. Especially since the so-called new synthesis of the early twentieth century, inheritance had tended to be understood primarily as the transmission of *genetic* material between generations and as largely divorced, conceptually, from developmental processes. Today this understanding has been drastically unsettled by the realization that it is not just genes (along with epigenetic factors) that are inherited in meaningful and vital ways. As Paul Griffiths and Russell Gray (2001) put it, the concept of inheritance ought to be applied "to any resource that is reliably present in successive generations, and is part of the explanation of why each generation resembles the last" (196). In this context, biologists are required to think about a range of factors that might be thought about as "environmental" (from the very particular developmental space of the womb to the larger ecosystem) and factors that might be called "cultural" (behaviors, languages, and more), as being in an important sense passed between generations, enabling the continuity of particular ways of life (Jablonka and Lamb 2005; Oyama, Griffiths, and Gray 2001).

Some of these inheritances are linear—from biological parent to offspring—but they are also more than this: they are radically multivalent and radically multispecies. Natural selection can operate on all these forms of inheritance, but (depending on how it is defined) it is rarely the only form of selection at work. Referencing the work of biologist Scott Gilbert (Gilbert, Sapp, and Tauber 2012), Donna Haraway (2014) has noted that we are all "lichens": beings composed as, and out of, entanglements of diverse others, shaped by inheritances much more complex than a genetic blueprint. From within the context of this emerging paradigm, the "cultural" and the "biological," the "evolutionary" and the "developmental," cannot be neatly teased apart. In Deborah Bird Rose's (2012) terms, life is a product of both sequential and synchronous relationships and inheritances. Who we all are as individuals, as cultures, as species, is in large part a product of generations of co-becoming in which we are woven through with traces of all of our multispecies ancestors.

An appreciation of these kinds of entanglements makes it easier to understand why a species like 'alalā cannot be neatly excised from our living world. Each species is a strand in a fabric, what I have elsewhere called a "flight way"—a term that aims to evoke an understanding of species as evolving ways of life, as interwoven lines of intergenerational movement through deep history. In this context, extinction always takes the form of an unraveling of co-formed and -forming ways of life, an unraveling that begins long before the death of the last individual and continues to ripple out long afterward: hosts of living beings—human and not—are drawn into extinctions as diverse heritages breakdown or are otherwise transformed (van Dooren 2014).

There is no solid line here between "human" and "ecological" dimensions, between evolutionary and cultural entanglements: relationships and affinities cut across any simple divide, moving back and forth with ease. The traces that we leave behind in one another remind us that conventional Western notions of "the human" as a being set apart from the rest of the living world have always been illusory (Plumwood 1993). In Anna Tsing's (2012) terms, "Human nature is an interspecies relationship"; it is the shifting historical product of "varied webs of interspecies dependence" (144). As it is sometimes succinctly put by Native Hawaiians: the people arrived as Polynesians, but the islands made them Hawaiian.

Spectral Crows and the Promise of Return

As I traveled, observed, and talked with a range of people on a research trip in Hawai'i, I encountered another important site in which the absence of crows was helping to shape future possibilities for everyone. At the center of this story is the Ka'ū Forest Reserve in the south of the Big Island—the forest in which I stood listening and hoping for crows. Early in my trip, I traveled high up into this area with a group of conservationists and state and federal land managers, a two-hour drive on a very bumpy dirt road that crossed old paddocks, forested areas, and cooled lava fields that stretched out black into the distance as far as the eye could see.

Just a few months earlier, the state government had released its management plan for the area. At the core of the plan was a proposal to fence 20 percent of the reserve, almost 12,355 acres (State of Hawai'i 2012). The fenced section would still allow human visitors, but all the pigs inside would be killed so that the understory might recover. Hopes and dreams for the future of 'alalā animated this proposal, at least in part. As the forest recovers, it is anticipated that it will be a future release site for these birds—while also contributing to the conservation of a range of other endangered species and ensuring that erosion is minimized so that the forest remains a healthy water catchment.

But not everyone supported this plan. Although its drafting involved more than a year of serious community consultation, it was greeted with hostility by some locals. The most vocal opposition came from hunters—some of them Native Hawaiians—who do not want to see a fence built and the pigs that they hunt removed from the area. Of course, hunters are a diverse crowd in most places, and this is certainly true in Hawai'i. In this context, opposition to fencing is grounded in a range of understandings, values, and histories. On the surface, the most prominent opposition to this fence has been justified by the notion that there is not enough accessible public hunting land in Hawai'i, while too much land is already "locked up" in conservation.[2] In short, for these people it is often simply a question of whether the interests of birds, snails, and plants should take priority over those of humans. In addition, hunters often challenge the notion that pigs and other ungulates damage the forest, some even arguing that pigs actually play a positive ecological role: tilling the soil and rooting out weeds.[3]

The three conservationists who led our little expedition to the Ka'ū Forest Reserve that day were all locals, born and raised in the district of Ka'ū. Both John, a former ranch hand, a longtime hunter, and a conservation convert, and Shalan, an ecologist, worked for the Nature Conservancy. Nohea, a young Hawaiian woman with deep family roots in the area and a degree in Hawaiian studies, was working as a community outreach and education officer for the state government. Together, they played a central role in the drafting of the new management plan for the area, especially the community-engagement process.

As part of this process, they took numerous groups of locals, including many hunters, up to the section of forest that the state is proposing to fence. After visiting the site, many hunters who were initially skeptical agreed that fencing is a good idea: partly because the visit impressed upon them just how remote the area is (and therefore inconvenient for hunting), but also because they were able to see with new eyes—with biologists' eyes, perhaps—the extent of the damage that ungulates were doing to the forest.[4]

During these site visits, John, Shalan, and Nohea also spent a lot of time talking to local people on the long drive up and back. John explained to me that one of the ways in which he conveyed the significance of the extinction of the 'alalā to local people was to draw a direct comparison between the loss of this species, on the one hand, and the potential loss of Hawaiian language and culture, on the other. The value of inherited diversities, of sustaining them into the future, was the point here. While John was mindful of the fact that cultural and linguistic diversity often rely on biodiversity (and vice versa) (Maffi 2004; Martin, Mincyte, and Münster 2012), his main point in making this connection in discussions with hunters was as a means of illustrating how biological "species" might themselves also be a kind of valuable diversity in our world. The tragedy of lost cultures in a colonized land allows people to connect with the loss of a bird, which, for some, had come to seem insignificant.[5]

These sites of communication and contestation between conservationists and hunters are from the outset about much more than 'alalā. The imagined and inherited past haunts the present in often unexpected ways. A key part of this haunting is the way in which the particular histories that we tell, that we inhabit, animate our understanding and action. Histories are not *of* the world, but *in* the world, as Haraway (2016:14) reminds us of stories in general. And so, how we tell the past, as well as which pasts we tell, plays a powerful role in structuring what is nurtured into the future and what is allowed or required to slip away. All the rich cultural and biological inheritances that constitute our world are at stake, to a greater of lesser extent, in the histories that we weave out of, and into, this forested landscape.

Of course, some hunters opted not to go on site visits to the Ka'ū Forest Reserve, and others remained unconvinced. Many of these people continue to oppose the fencing and removal of pigs from this area; some of the most vocal opponents are a small group of Native Hawaiian hunters. For many Native Hawaiians, pig hunting is understood as a core traditional practice that ought to be widely supported as part of the continuity of Hawaiian culture. In conversations with these hunters, as well as in online discussion forums, I encountered repeated references to this point of view. For them, any effort to remove pigs and limit hunting is seen as a violation of their traditional and customary rights, protected by the Hawaiian constitution (sec. 7).[6]

In recent years, however, the notion that pig hunting is a traditional cultural practice has been thoroughly problematized. Detailed historical studies by Hawaiian cultural experts Kepa and Onaona Maly indicate that prior to European arrival, pigs were kept close to home, and they were also distinctly different animals: the smaller Polynesian variety, not like the large European boars now found widely throughout the islands. The only hunting that likely took place at that time was bird hunting, primarily for feathers used in royal ornaments and clothing (Maly and Maly 2004:152; see also Gon, n.d.; Maly, Pang, and Burrows 2007).

With this information fresh in my mind, I expected conservationists to readily dismiss claims by hunters to "tradition," but found that this was not the case. Instead, almost all the conservationists I met with noted that this shorter history did not invalidate claims to continued hunting. Many noted that the length of time required to make something "traditional" was uncertain, that culture is not static, and that several generations of hunting is certainly long enough to establish family traditions—forms of identity and culture— that ought to be respected wherever possible. In short, they recognized in their own way that, as James Clifford (1986) has famously put it: "'Cultures' do not hold still for their portraits" (10).

But something else was happening here, too. Several of the conservationists that I spoke with quickly mentioned this historical research when the topic of pig hunting came up. Although they were clear that this did not mean that hunters had no claim to continue hunting, it clearly changed the *nature* of that claim. In noting that the pigs are different from those originally brought to the islands by Polynesians and the practice is more recent than sometimes thought, a break with the past is effected in which fencing and pig removal are conceptually separated from contentious questions of Native Hawaiian customary practice and rights. Different histories create different continuities and ruptures, with all their attendant political and ethical consequences (Bastian 2013). Importantly, however, it was not just *haole* (white) conservationists making this claim; in fact, some of the people who made it most strongly to me in interviews were Native Hawaiians who see the removal of pigs from at least some areas of forest as essential to the conservation not only of the environment, but of a rich notion of Hawaiian culture, too. I will return to this topic.

The desire of some conservationists to conceptually separate pig hunting from traditional Hawaiian culture is, I believe, in large part an effort to *depoliticize* plans to remove pigs. This is nowhere more clear than in the prominent role that the history of the occupation of Hawai'i by the United States is playing in some of the most vocal opposition to fencing in Ka'ū. With the occupation firmly in mind, for some hunters the proposed fence is one more "land grab" in a long history of taking.

The last monarch of the sovereign nation of Hawai'i, Queen Lili'uokalani, was overthrown in 1893 by a group of wealthy settlers with the aid and support of members of the United States government and its military. Through a complex series of events over the next five years, Hawai'i became a territory of the United States and fifty years later was made a state. Although there was some attempt, both in the lead up to the overthrow and afterward, to provide Native Hawaiian commoners with some form of property rights in small parcels of land, this never really worked out in their favor (Banner 2007; Silva 2004): from the Great Mahele of 1848, through subsequent decades of dispossession and annexation, until, in J. Kehaulani Kauanui's (2008) words, "by the mid-nineteenth century, Hawaiians and their descendants [had become] largely a landless people" (75).[7]

For people inhabiting this history, fence building is never an innocent act. In this context, conservation is regarded as one more excuse to take away people's rights to access or use land. As one hunter put it, environmentalists are "always using something endangered to the i[s]lands for try grabb land."[8] Importantly, these people do not trust the intentions of government agencies in this area, viewing any fencing as the beginning of a slippery slope toward complete loss of access. As another hunter put it: "Environmentalist want to eventually take it all away and fence it in! They're starting with these areas, and will start working on more. The alala, water shed, native plants, etc. is just a smoke screen to grab more land!"[9]

There is something very familiar about these views. In many parts of the world—including the U.S. mainland—hunters express similar concerns about conservation (Emery and Pierce 2005; McCarthy 2002). But there is also something distinctly Hawaiian about them; there are clear echoes of the Great Mahele and acts of subsequent dispossession here, as well as frequent references or allusions to traditional rights. Perhaps most important, however, these arguments by hunters often explicitly challenge the authority of the state government and, certainly, that of the federal government—illegal governments from this perspective—to exercise any authority in the management of these lands and resources.

This connection between conservation and occupation does important political work. Once a proposal like the Kaʻū Forest Reserve Management Plan has been framed by critics in this way, those who speak in its favor are positioned as endorsing the occupation. As Shalan Crysdale put it to me in an interview: "To be for the plan is to be for the overthrow."[10] In this context, publicly supporting conservation—as a Hawaiian or anyone else—requires one to enter into what another local called the "raging fire of emotion" that surrounds the occupation and subsequent colonization of the islands.

In this light, 'alalā themselves become an enemy of the Hawaiian people. What's more, the birds' movements through the forest become suspect as hunters fear that each time 'alalā move beyond the fenced area (especially if they are nesting), the fence will expand with them. And so, the 'alalā is imagined as a Trojan horse of sorts whose conservation facilitates further loss of land and rights. It should come as no surprise that in this climate, conservationists have real fears that any released birds will be targeted by some hunters.

INHERITING THE WORLD

Toward the end of my most recent trip to Hawai'i, I met with Hannah Kihalani Springer, a *kupuna*, or elder, who lives in the district of North Kona. She is deeply knowledgeable about Hawaiian history and culture, about hunting and conservation, so I was eager to hear her thoughts on the past and future of the islands. Sitting in her living room in her family's old homestead, we talked about conservation, politics, sovereignty, ranching, and, of course, 'alalā. Hannah is lucky enough to have seen free-living 'alalā, sometimes in large gangs, throughout her early life. She recalled:

When we went into certain sections of our lands, it was the norm to see crows. From the '50s, all the way through to the mid-'70s ... It was January 1, 1977—and I only know the date because my mother's birthday was January 1, and I had gone to pick maile [a plant] for her birthday—when two adults and a young bird came and worried me. That was the last close up encounter that I had with wild crow.

Hannah is a passionate and active conservationist, president of the Conservation Council for Hawai'i. Like many other people with whom I spoke, she feels that in some places pigs and other ungulates need to be fenced out and removed for conservation. But she also believes that room has to be made for hunters—her family hunts, and in the past she hunted too. And so, like others I spoke with, she feels that the government could do more to facilitate access to existing state land for hunting.

In contrast to those Hawaiians who strongly emphasize the place of pig hunting in their culture, Hannah noted that the islands' forests are alive with a diversity of plants and animals, all of which have their places in Hawaiian stories and culture. In this context, she argued that a singular focus on pigs is not helpful. In her words: we need "the larger context that is much more diverse and dynamic. . . . When we so diminish the conversation we're diminishing the Hawaiian experience and the Hawaiian culture. The forest is important for the myriad characteristics that comprise the whole."

Other Hawaiians that I spoke with who share this view referenced another history—the Kumulipo, an origin story—in their arguments about the need to hold onto a diversity of plants and animals in the forest. For these people, removing pigs from portions of the forest to aid in the conservation of 'alalā, other endangered birds and plants, and the watershed is essential for the protection of Hawaiian life and culture. This is perhaps particularly the case in a place like the Ka'ū Forest Reserve, where, even if this fence did go ahead, the remaining 80 percent of the area would still be open to pigs and hunters.

Speaking with Hannah that day, I was reminded again and again that the histories that we tell are themselves *acts* of inheritance. Which is to say, that the aspects of the world that we nurture into the future are, in more or less significant ways, shaped by how we understand and tell the past. Histories structure our understandings of what particular continuities mean and why they matter.

There is an important dynamic at work in inheritance here that deserves further attention. In *For What Tomorrow . . . A Dialogue* (2004), Jacques Derrida excavates the basic structure of inheritance. He is primarily interested in what it means to inherit traditions, languages, and cultures. At its simplest level, inheritance seems to be about continuity and retention: taking up the past and carrying it forward into the future. Of course, much of this inheritance is not actively chosen: we are thrown into our heritage; it "violently elects us." But this is not the end of the story. For Derrida, in any act of inheritance there is also transformation. While language, culture, and tradition all continue from generation to generation, they are living heritages not fixed once and for all. It is this "double injunction" at the heart of inheritance that Derrida draws attention to, describing the act of inheritance as one of "reaffirmation, which both continues and interrupts" (Derrida and Roudinesco 2004:4).

But this dynamic extends well beyond the human domains that so interest Derrida. All living beings are involved in their own forms of life- and world-shaping inheritance, *which include both retention and transformation.* Evolution by natural selection—that great engine of new ways of life—is grounded in forms of inheritance that simultaneously retain the achievements of the past while constantly transforming them to produce new variability. This variability arises through recombination, mutation, and other forms of transformation, and is the stuff of future change and adaptation. Moving beyond the narrow genetic reductionism commonly found in neo-

Darwinian accounts, we are reminded that these are lively and varied processes in which diverse heritages move between organisms in a range of different ways to shape bodies and worlds.[11]

In this context, the fundamental structure of life is one of inheritance. Darwin knew something like this when he drew a comparison between language and biological species, with an emphasis on the way in which both are at their core *genealogical*: seemingly "individual" languages and "individual" species are in reality simply moments within longer historical lineages (Grosz 2004). Here, life takes shape through the constant generation of variability, only some of which "sticks," only some of which is retained and so incorporated into the larger collective (be it a language, a species, or indeed a culture). As Derrida succinctly put it: "Life—being alive—is perhaps defined at bottom by this tension internal to a heritage, by this reinterpretation of what is given" (Derrida and Roudinesco 2004:3–4).[12]

Inheritance is a productive concept for extinction studies and the broader environmental humanities; a concept with a long and rich history in both the biological and the human sciences. Reading Derrida with Darwin—or, better yet, with more recent work in evolutionary and developmental biology, developmental systems theory, and related fields—we are able to begin to develop an appreciation for entangled *biocultural* inheritances in which the movements of genes, ideas, practices, and words between and among generations cannot be isolated into separate channels of inheritance.[13] If we scratch the surface just a little, these entanglements are palpable in Hawai'i's shrinking forests: as the island's biotic diversity continues its long role in helping to nourish and shape local cultures, cultures that are, in turn, remaking those ecologies and the futures of their many inhabitants.

Thinking in this entangled way draws us, inexorably, into an understanding of the ethical work of inheritance. Where species, ecologies, and cultures are in processes of ongoing and dynamic change, much of what is and is not passed on is not up to any of us. Where we can and do play a role, however, the question is usually the same. Never simple, never clean: *What* is to be lost and what retained? Which losses will we accept, and in the name of which continuities (and vice versa)? From within a time of colonization and extinction—a time in which so much of this biocultural diversity is being lost, often violently—what does it mean to inherit responsibly, and how might we live up to our inheritances?[14]

One of the many things that I learned from Hannah was the fact that responsible inheritance is necessarily grounded in a recognition of, and an attentiveness to, multiple voices, with their diverse histories and imagined futures.[15] As our conversation was coming to a close that afternoon, Hannah and I drifted into a discussion of the sovereignty movement in the islands. She told me about a relative of hers, deeply committed to Hawaiian sovereignty, who worked for the state government as a biologist. When asked about the incompatibility between her politics and her employment, this relative would say that she was conserving Hawai'i's biotic diversity so that when and if sovereignty comes, the people and the land are in the best possible condition for it. Although Hannah didn't explicitly state it, it seemed to me that she herself shared this general view. She went on to say:

> The conclusion that I've arrived at is: "I am a citizen of the land." We have lived on this land, as I've described to you, since before Cook's arrival. And, we've seen chiefs rise and fall, we've seen an island nation born and die before its time, elected and appointed officials come and go, but here we stand. I'm less interested in the constitution that binds us or the flag that flies over the land, than I am in the quality of life on the land. So, if there are elements within whoever's constitution it is that allow us to preserve and pursue the righteous management of the resources that we call home, then I am happy to

pursue those. . . . I am loyal to this land. Whatever flag flies over it is one that I am willing to use the resources of to continue to be a citizen of this land.

Hannah's position is one of hope, within which resides a profound responsibility to both the past and the future. Hannah has not forgotten the events of 1893. But she wants to inherit this history in a way that refuses to regard support for conservation as necessarily support for an illegal occupation. She wants to inhabit the history of these islands, her and her family's history, in a way that holds open possibilities for flourishing life—for the landscape and the people who are a part of it—into the distant future. In short, she is proposing that we might care for ʻalalā, *and* for Hawaiian culture and sovereignty, *and* for the rest of the land and its people.[16]

Of course, there will always be compromises and challenges here, and they will likely always be unequally distributed. But I am inspired by Hannah's effort not to abandon any of these inheritances, to pay attention to their entanglements, and to take on the work of nourishing them as a responsibility to the past and the future to come.

Here, I think we see that responsible inheritance requires that we engage with others—their histories, their relationships—to hold open a future that does not forget the past or attempt to reconstruct it, but rather inherits it as a dynamic and changing gift that must be lived up to for the good of all those who do or might inhabit it. This is what Rose (2004) has called "recuperative work," work that begins from the conviction that, in her words:

there is no former time/space of wholeness to which we might return or which we might resurrect for ourselves. . . . Nor is there a posited future wholeness which may yet save us. Rather, the work of recuperation seeks glimpses of illumination, and aims toward engagement and disclosure. The method works as an alternative both to methods of closure or suspicion and to methods of proposed salvation. (24)

In this context, "taking care" is always a historical and a relational proposition; if we're doing it right, care always thrusts us into an encounter with ghosts, our own *and others'*. Some people live in worlds haunted by evolutionary ghosts: anachronistic plants and lost seed dispersers. Others live in worlds haunted by the wrongs of 1893 and dreams of a sovereignty to come. Others remember 'alalā in the forest when they were children, or are tied to *this* bit of forest by memories of a grandfather who taught them to hunt. Responsibility resides in a genuine openness to these diverse voices with all their complex pasts and futures.

But, importantly, care and responsibility necessarily draw us out beyond the arbitrary and unworkable limits of a purely human space of inheritance and meaning making. In short, "ours" aren't the only hauntings that constitute worlds. Some plants live and are now disappearing in worlds haunted by 'alalā; some crows are drawn, *called*, to a forest beyond the aviary. Paying attention to diverse voices means recognizing that nonhumans are not simply resources to be conserved or abandoned, inherited or cast aside, on the basis of whether current generations of humans happen to want them around. Rather, 'alalā, ho'awa, and others are themselves constituted through immense processes of intergenerational life, the cumulative achievement of multispecies entanglements, adaptation, and inheritance across vast periods of time. As such, their own ongoing dramas as well as those of the many other forms of life that have already made, and might yet still make, worlds with them demand our respect and gratitude (van Dooren 2014).

In paying attention to some of the diverse ways that nonhumans inherit their worlds, we become aware of just how much is at stake in extinction. For example, there are now suggestions that in captivity, the once remarkable vocal repertoire of 'alalā—their raucous calls and mournful songs—is being diminished. Perhaps this is because they have less to talk about, or perhaps juvenile birds simply haven't been exposed to enough chatter from their elders. Similarly, know-how about predators and how to avoid them may not be being passed between generations in captivity, potentially greatly affecting their future survival (van Dooren 2016). In these and other ways, the long-accumulated heritage of the species—not just its genetics, but learned behaviors that took advantage of generations of refinement and adaptation—are now perhaps being undermined, to the detriment of any future life for 'alalā in the forest (and despite great effort by their human carers). Here we see in the most tragic of ways that as a species, and as individual birds, 'alalā are historical beings with their own inheritances. Much is at stake *for* them, not just *in* them at the edge of extinction. Furthermore, as we are seeing, the histories that humans tell play a significant role in shaping whether, and in what ways, 'alalā are able to take up these heritages to contribute to the crafting of vibrant and thriving worlds for themselves and others.

Ours is a time of mass extinction, a time of ongoing colonization of diverse human and nonhuman lives. But it is also a time that holds the promise of many fragile forms of decolonization and hopes for a lasting environmental justice. Here, the work of holding open the future and responsibly inheriting the past requires new forms of attentiveness to *biocultural* diversities and their many ghosts. But beyond simply listening, it also requires that we take on the fraught work—never finished, never innocent—of weaving new stories out of this multiplicity. Stories within stories that bring together the diversity of voices necessary to responsibly inhabit the rich patterns of interwoven inheritance that constitute our world.

ACKNOWLEDGMENTS

This research was made possible by funding from the Australian Research Council (DP110102886; DP150103232). I would also like to thank the many people in Hawai'i who gave their time to discuss 'alalā and the pasts and futures of conservation in the islands—in particular, Hannah Kihalani Springer, John Replogle, Shalan Crysdale, Nohea Ka'awa, Rich Switzer, Alan Lieberman, Paul Banko, and Donna Ball. This chapter was originally presented as the Sir Keith Hancock Lecture to the Forty-fourth Annual Symposium of the Australian Academy of the Humanities.

NOTES

1. Jeff Burgett (agent with the U.S. Fish and Wildlife Service), interview with the author, Hilo, Hawai'i, December 19, 2011.

2. Private land is one of the key obstacles here. In some cases, privately owned lands are being closed off to hunters (perhaps because of insurance concerns or landowners' bad past experiences with hunters). In other cases, public land where people might hunt is inaccessible because the owners of private properties surrounding it—often remnants of large plantations or ranches—restrict direct or open access to it. In addition, it should be noted that relatively little state land is actually utilized solely (or even primarily) for conservation purposes (Lisa Hadway, interview with the author, January 25, 2013. Hadway is the manager of the state government's Natural Area Reserves System, Division of Forestry and Wildlife, Department of Land and Natural Resources). At present, the Division of Forestry and Wildlife (DOFAW) provides 600,000 acres of public hunting area on the island of Hawai'i. Of this land, "only about 4 percent is currently fenced with hooved animal populations effectively controlled [a requirement for effective conservation]. Under the most ambitious current plans for fencing and

ungulate removal over the next decade, about 17 percent of DOFAW lands on the island would be affected, most of which would occur on Mauna Kea" (Geometrician Associates 2012:86).

3. Anonymous interviewees. Unless otherwise noted, these interviews were conducted by the author with biologists, managers, hunters, Native Hawaiians, and other locals in January and February 2013 on the islands of Hawai'i and O'ahu. In most cases, I have identified participants by name; in a few cases, where more appropriate, I have referenced them anonymously.

4. What counts as "damage" is a complex question. In large part, it is precisely this question that this chapter seeks to address. I am not of the view that some prior wilderness state or "natural balance" marks the way that the forest ought to be. Rather, the questions are precisely which kinds of forests we are trying to achieve, what values and goals ought to underlie our actions in forests, and how might we take a diverse range of human and nonhuman voices seriously in these discussions. Asking these questions is about undermining the obviousness of any assumed goals for forest ecosystems; it is about being specific about the values that guide understanding and action to shape worlds.

5. I accept J. Kehaulani Kauanui's (2009) argument about the appropriateness of the term "colonization" to describe the social and political dynamics of Hawaiian life after what was technically an "occupation'" of the internationally recognized sovereign nation of Hawai'i. See also Silva (2004).

6. This comment was either made directly to me or presented by others as a claim commonly made, in several anonymous interviews conducted in January 2013. Similar comments can be found posted to the "hunting forum" Hawaii Sportsman, http://hawaiisportsman.forumotion .com/t5382-big-island-video-news-hunters http://hawaiisportsman.

7. The Great Mahele was a period of land redistribution—initiated by the king and the parliament of Hawai'i—that "converted" traditional customary rights in lands into private property in the mid-nineteenth century (in the lead-up to U.S. occupation) (Banner 2007; Silva 2004).

8. "Blue Mountain Traila," comment posted to Hawaii Sportsman, June 6. 2012, http://hawaiisportsman.forumotion.com/t5382-big-is land-video-news-hunters.

9. "Shrek," comment posted to Hawaii Sportsman, June 9, 2012, http://hawaiisportsman.forumotion.com/t5382p15-big-island-video -news-hunters. There does seem to be something to these arguments. Interviews that I conducted with conservationists, alongside their own public submissions during the community consultation process for the Kaʻū Forest Reserve Management Plan, make clear that most of them see protecting only 20 percent of the area as, in effect, sacrificing 80 percent. Many of them would like to see a lot more of the area fenced and ungulates removed. It is unclear exactly where the state stands on this, especially in the long term. Its position seems usually to involve some sort of middle ground that leaves both sides equally unhappy.

10. Shalan Crysdale (ecologist with the Nature Conservancy), inter- view with the author, Naʻalehu, Kaʻū District, Hawaiʻi, February 7, 2013.

11. This position should not be taken to imply that natural selection (however broadly defined) represents the only form of generating novelty, or an exhaustive explanation for the diversity of life. For an evocative and creative account of some of the many ways that life, and indeed evolution, exceed natural selection, see Hustak and Myers (2012).

12. Derrida seems to be thinking here about "Life" in a narrower sense than I am, with quite a tight focus on tradition, culture, and lan- guage (in human and, in particular, philosophical contexts).

13. The capacity to tell these stories about inheritance is, of course, *itself* a part of what we inherit from those who have come before us. The cognitive capacities, the cultural traditions (including those of evolu- tionary theory and the broader natural sciences), that make this awareness possible are themselves gifted to us within and by a historical world. Of course, the capacity to care about any of this is also a part of this heritage (van Dooren 2014:32–43).

14. Derrida's primary concern in his discussion of responsibility and inheritance is political conservatism and those modes of inheritance that

uncritically take up and perpetuate the past. In this context, responsibility emerges as a radical questioning of what is to be retained and what lost or transformed. In Derrida's terms, it is only through "reinterpretation, critique, displacement, that is, an active intervention, . . . that a transformation worthy of the name might take place; so that something might happen, an event, *some* history, an unforeseeable future-to come" (Derrida and Roudinesco 2004:4). The basic point here is simple and powerful. Inheritance that is mere repetition closes off the future, or rather, closes off the possibility of anything genuinely different and maybe, just maybe, better. Thanks to Rosalyn Diprose (2006) for her reading of Derrida and for being willing to chat about responsibility and inheritance with me. For a fuller discussion of Derrida's notion of a responsibility "worthy of the name," see Diprose (2006).

15. I have no particular authority to speak on this matter in Hawai'i. But I am drawn by a genuine concern for the future of these forest and all their inhabitants to attempt to weave my way through these difficult topics, to arrive at some sense of "where to from here." Ultimately, however, I do not intend to argue for the "right to an opinion" on this topic. This essay is written in large part against the proposition that some people might be shut out of conversations that aim to imagine what responsibility and justice might look like in multispecies and multicultural worlds, solely on the basis of the kinds of inheritance that they bring with them, that they don't have the right kinds of history. Furthermore, from my perspective, the relevant ethical obligation is a demand issued on all sentient creatures to respond when they are witness to suffering, violence, and death.

16. While we often place great emphasis on the past in discussions of inheritance, the ethical work that it demands is equally oriented toward imagined futures in which traces of the then past will *matter* in some way. In short, inheritance is also a question of what we will leave behind for those yet to come.

REFERENCES

Banner, Stuart. 2007. "Hawaii: Preparing to be Colonized." In *Possessing the Pacific: Land, Settlers, and Indigenous People from Australia to Alaska*, 128–162. Cambridge, Mass.: Harvard University Press.

Barad, Karen. 2007. *Meeting the Universe Halfway: Quantum Physics and the Entanglement of Matter and Meaning*. Durham, N.C.: Duke University Press.

Barlow, Connie. 2000. *The Ghosts of Evolution: Nonsensical Fruit, Missing Partners, and Other Ecological Anachronisms*. New York: Basic Books.

Bastian, Michelle. 2013. "Political Apologies and the Question of a 'Shared Time' in the Australian Context." *Theory, Culture & Society* 30, no. 5:94–121.

Clifford, James. 1986. "Introduction: Partial Truths." In *Writing Culture: The Poetics and Politics of Ethnography*, edited by James Clifford and George E. Marcus, 1–26. Berkeley: University of California Press.

Culliney, Susan Moana. 2011. "Seed Dispersal by the Critically Endangered Alala (*Corvus hawaiiensis*) and Integrating Community Values into Alala (*Corvus hawaiiensis*) Recovery." Master's thesis, Colorado State University.

Culliney, Susan, Liba Pejchar, Richard Switzer, and Viviana Ruiz-Gutierrez. 2012. "Seed Dispersal by a Captive Corvid: The Role of the 'Alalā (*Corvus Hawaiiensis*) in Shaping Hawai'i's Plant Communities." *Ecological Applications* 22, no. 6:1718–1732.

Derrida, Jacques, and Elisabeth Roudinesco. 2004. *For What Tomorrow . . . A Dialogue*. Translated by Jeff Fort. Stanford, Calif.: Stanford University Press.

Diprose, Rosalyn. 2006. "Derrida and the Extraordinary Responsibility of Inheriting the Future-to-come." *Social Semiotics* 16, no. 3:435–447.

Emery, Maria R., and Alan R. Pierce. 2005. "Interrupting the Telos: Locating Subsistence in Contemporary US Forests." *Environment and Planning A* 37:981–993.

Geometrician Associates. 2012. *Final Environmental Assessment—Ka'u Forest Reserve Management Plan*. Honolulu: Prepared for State of Hawai'i, Department of Land and Natural Resources.

Gilbert, Scott F., Jan Sapp, and Alfred I. Tauber. 2012. "A Symbiotic View of Life: We Have Never Been Individuals." *Quarterly Review of Biology* 87, no. 4:325–341.

Gon, Sam 'Ohukani'ohi'a, III. n.d. "Pua'a: Hawaiian Animal—or Forest Pest?" Unpublished manuscript, on file with author.

Griffiths, Paul E., and Russell D. Gray. 2001. "Darwinism and Developmental Systems." In *Cycles of Contingency: Developmental Systems and Evolution*, edited by Susan Oyama, Paul E. Griffiths, and Russell D. Gray, 195–218. Cambridge, Mass.: MIT Press.

Grosz, Elizabeth. 2004. *The Nick of Time: Politics, Evolution, and the Untimely*. Durham, N.C.: Duke University Press.

Haraway, Donna. 2014. "SF: String Figures, Multispecies Muddles, Staying with the Trouble." Lecture presented at the University of Alberta, Edmonton, March 24.

——. 2016. *Staying with the Trouble: Making Kin in the Chthulucene*. Durham, N.C.: Duke University Press, 2016.

Hustak, Carla, and Natasha Myers. 2012. "Involutionary Momentum: Affective Ecologies and the Sciences of Plant/Insect Encounters." *Differences* 23, no. 3:74–118.

Jablonka, Eva, and Marion J. Lamb. 2005. *Evolution in Four Dimensions: Genetic, Epigenetic, Behavioral, and Symbolic Variation in the History of Life*. Cambridge, Mass.: MIT Press.

Janzen, Daniel H., and Paul S. Martin. 1982. "Neotropical Anachronisms: The Fruits the Gomphotheres Ate." *Science* 215:19–27.

Kauanui, J. Kēhaulani. 2008. *Hawaiian Blood: Colonialism and the Politics of Sovereignty and Indigeneity*. Durham, N.C.: Duke University Press.

——. 2009. "Hawaiian Independence and International Law." Episode 23, November 23. Indigenous Politics: From Native New England and Beyond. http://www.indigenouspolitics.org/audio-files/2009/11-23%20Hawaiian%20Independence.mp3.

Leonard, David L., Jr. 2008. "Recovery Expenditures for Birds Listed Under the US Endangered Species Act: The Disparity Between Mainland and Hawaiian Taxa." *Biological Conservation* 141:2054–2061.

Maffi, Luisa. 2004. "Maintaining and Restoring Biocultural Diversity: The Evolution of a Role for Ethnobiology." In *Ethnobotany and Conservation of Biocultural Diversity*, edited by Thomas J. S. Carlson and Luisa Maffi, 9–35. New York: New York Botanical Garden Press.

Maly, Kepā, and Onaona Maly. 2004. *He Moʻolelo ʻĀina: A Cultural Study of the Manukā Natural Area Reserve Lands of Manukā, District of Kaʻū, and Kaulanamauna, District of Kona, Island of Hawaiʻi.* Hilo: Kumu Pono Associates, for State of Hawaiʻi, Department of Land and Natural Resources.

Maly, Kepā, Benton Kealiʻi Pang, and Charles Peʻapeʻa Makawalu Burrows. 2007. "Pigs in Hawaiʻi, from Traditional to Modern." Unpublished manuscript, on file with author.

Martin, Gary, Diana Mincyte, and Ursula Münster, eds. 2012. "Why Do We Value Diversity? Biocultural Diversity in a Global Context." *RCC Perspectives: Transformations in Environment and Society*, no. 9.

McCarthy, James. 2002. "First World Political Ecology: Lessons from the Wise Use Movement." *Environment and Planning A* 34:1281–1302.

Munro, George. 1944. *Birds of Hawaii.* Honolulu: Tongg.

Oyama, Susan, Paul E. Griffiths, and Russell D. Gray, eds. 2001. *Cycles of Contingency: Developmental Systems and Evolution.* Cambridge, Mass.: MIT Press.

Plumwood, Val. 1993. *Feminism and the Mastery of Nature.* New York: Routledge.

Rose, Deborah Bird. 2004. *Reports from a Wild Country: Ethics for Decolonisation.* Sydney: UNSW Press.

——. 2012. "Multispecies Knots of Ethical Time." *Environmental Philosophy* 9, no. 1:127–140.

Silva, Noenoe K. 2004. *Aloha Betrayed: Native Hawaiian Resistance to American Colonialism.* Durham, N.C.: Duke University Press.

Sodikoff, Genese Marie. 2013. "The Time of Living Dead Species: Extinction Debt and Futurity in Madagascar." In *Debt: Ethics, the Environment, and the Economy*, edited by Peter Y. Paik and Merry Wiesner-Hanks, 140–163. Bloomington: Indiana University Press.

State of Hawai'i. 2012. *Ka'ū Forest Reserve Management Plan*. Honolulu: State of Hawai'i, Department of Land and Natural Resources, Division of Forestry and Wildlife.

Steadman, David W. 2006. *Extinction and Biogeography of Tropical Pacific Birds*. Chicago: University of Chicago Press.

Tsing, Anna Lowenhaupt. 2012. "Unruly Edges: Mushrooms as Companion Species." *Environmental Humanities* 1:141–154.

van Dooren, Thom. 2014. *Flight Ways: Life and Loss at the Edge of Extinction*. New York: Columbia University Press.

——. 2016. "Authentic Crows: Identity, Captivity and Emergent Forms of Life." *Theory, Culture & Society* 33, no. 2:29–52.

Hayashi and Toda, "Male Passenger Pigeon (*Ectopistes migratorius*)." (From Charles Otis Whitman, *Orthogenetic Evolution in the Pigeons* [Washington, D.C.: Carnegie Institution of Washington, 1920])

AFTERWORD

It Is an Entire World That Has Disappeared

VINCIANE DESPRET TRANSLATED BY MATTHEW CHRULEW

C'est un monde qui s'en va.

BANG!

September 1899, Babcock, Wisconsin. The last American Passenger Pigeon (*Ectopistes migratorius*) in the wild is shot by the last American hunter of Passenger Pigeons. Some say, however, that at least one more remained. It would be captured the following March. It didn't survive.

September 1, 1914, 1:00 P.M., Cincinnati Zoo, Ohio. Martha, the last female, miraculously preserved in captivity until then, passed away on the floor of her cage. She was twenty-nine years old. Her companion, George, had died four years earlier. The two had been the species's last chance. They declined. They preferred not to leave any descendants behind.

I imagine that she, Martha, must have closed her eyes, tranquil. She completed her first migration, and the last for all those whose existence she prolonged for several years. Or what is called an existence—a long moment of abstraction, a skyless existence. A bad existential gamble. Martha ceased to exist in a world that was no longer as it had once been. Let the world go on without us. She rejoined her partner and their kind. Let this whole story end . . .

Martha did not lay the white egg that could have prolonged this story. Neither she nor he wanted to hatch and then feed the little being who would have emerged—during those fifteen days that give rhythm to all parents' lives in the world of Passenger Pigeons, as it had for their parents and their parents' parents and theirs, so much so that they had populated the earth, trees, and sky, as no other bird had done until then. They could not have done it. As they could not have, at the end of those two weeks, according to the tried-and-true customs of such migratory birds, abandoned the little one, already quite plump, alone in the nest . . . moved away, and let it cry. With full confidence that it would learn on its own, grow up, and dare to let itself fall. And discover, on its own, the joyous need to fly. What sense would all of this make in a cage?

They could not or did not want to. They didn't want to start all over again, to start again from nothing, especially when nothing is *nothing*, an existence without others, an existence without sky.

Perhaps they still had, in the depths of a memory of which animals have the secret and that they transmit without our knowing, the memory of massacres, rifles, and trees that people set in flames in the darkest of night? Did they have an intuition of what was and what will have been? That the sky had become a desert? That to be ten, or even a hundred, means to be alone when you are a Passenger Pigeon? Did they know, from their ancestors' memories, that the land, forests, and fields, seen by few eyes, no longer resembled anything, and that their patterns and colors, so familiar and recognizable when the eyes are many, had become incomprehensibly foreign and senseless for theirs—like a painting by an artist gone mad? Not to mention the silence and everything that marked the absence: the triumphant fluttering takeoff of all the attuned bodies, the nights in the branches that creak in panic and the thunderous awakenings.

If only this story would end, it doesn't deserve to be prolonged . . .

In 1947, in a Wisconsin park, a monument was unveiled that commemorated the extinction of the migratory pigeon. This monument, Aldo Leopold (1949) wrote, "symbolizes our sorrow. We grieve because no living man will see again the onrushing phalanx of victorious birds, sweeping a path for spring across the March skies, chasing the defeated winter from all the woods and prairies of Wisconsin" (108–109).

"The monument's action is not memory but fabulation" (Deleuze and Guattari 1994:168).

Certainly, humanity has lost the presence of the birds. They have kept the names in memory but have forgotten what these names evoke—because the names render one sensitive to what they designate, but what few people know: Ectopistes migratorius—people have forgotten the freest voyages that life had ever invented. "Passenger Pigeon," the "pigeon that passes by": people will henceforth miss the surprise of the birds that are only passing by, not in a predictable seasonal rhythm, but at the mercy of the gifts that Earth offers them. Tourte voyageuse (Passenger Pigeon), tourte (pie), tourtière (meat pie): humans have erased the taste of this dish from their tongue, a dish that nourished them and for which they failed to give thanks. Colombe voyageuse, a dove colored like a rainbow, a black dove in an undulating multitude: it is said that long ago, before the massacres, when the pigeons passed through the sky, the swarm was so vast and dense that it created dark clouds like the ones that precede thunderstorms, and that the sun would disappear, sometimes for hours, sometimes for entire days. Humanity has lost winged eclipses.

But what the world has lost is not what people mourn.

What the world has lost, and what truly matters, is a part of what invents and maintains it as world. The world dies from each absence; the world bursts from absence. For the universe, as the great and good philosophers have said, the entire universe thinks and feels itself, and each being matters in the fabric of its sensations. Every sensation of every being of the world is a mode through which the world lives and feels itself, and through which it exists. And every sensation of every being of the world causes all the beings of the world to feel and think themselves differently. When a being is no more, the world narrows all of a sudden, and a part of reality collapses. Each time an existence disappears, it is a piece of the universe of sensations that fades away.

FLAP-FLAP-FLAP-FLAP.

The world had, with the Passenger Pigeons, the sensation of wings by the thousands. And without them, the wind—which had greatly contributed to their very invention—finds itself to be somewhat aimless, as do the updrafts and downdrafts, and the fresh breezes, tepid and hot, and the waves that are part of the journey, and the rays of light that took pleasure in shimmering, and the rustling trees shaken in all directions, and all of nature that was shadowed by their passage. And given this, I don't know who will rediscover the words, with these lost sensations, to describe the nostalgia of the sun, the one who had learned, with these great clouds of wings, to play hide and seek with the earth. The sun now has only some thunderstorms left, and—but in too rare and partial a way to rely on it—the moon. The world has lost the mischievous and untimely reinvention of darkness.

FLAP-FLAP-FLAP-FLAP-FLAP-FLAP-FLAP.

But what the world has lost even more is the unique, sensual, living, warm, musical, and colorful point of view that the Passenger Pigeons created upon it and with it. This unique point of view, to which the world owed the sensation of so many things, is no more. The happiness of being an immense wing traversing infinite spaces; the feeling of being a cloud above Earth and of creating changing shapes on it, flowing and shadowy: the sensation of the fields and the woods that, far below, fly by like the images of an accelerating film. The joy of being innumerable and of forming one perfectly attuned being, and the trust in this attunement, which is the figure of joy that the Passenger Pigeons invented when they learned to rely on the air and the wind. The world has lost the taste of dry and fleshy fruits, of seeds and insects, the raindrops that slide off feathers, the air that dances and that shapes the paths of heat and density, the music in the throbbing murmur of thousands of wings applauding the flight, the creaking of trees and branches shaken under the weight of rest, the shimmer of a rainbow that sweeps in search of the horizon . . . The perception of the vastness, of the innocence [*blancheur*] of an egg, and of the cry of a little one who feels itself abandoned.

All of this is no more. Humanity mourns the Passenger Pigeons. They also say that they should have been concerned, especially when they saw that as they passed in the sky, the sun continued to shine. Humanity can mourn the Passenger Pigeon. But it is the world that bursts with its absence.

NOTE

This piece was originally written as "P is for Passenger Pigeon" for Antonia Baehr and Friends, *ABeCedarium Bestiarium: Portraits of Affinities in Animal Metaphors* (Berlin: far° festival des arts vivants & make up productions, 2014). *ABeCedarium Bestiarium* was created by Baehr, a choreographer, as dance miniatures based on extinct animals. For further information, see http://www.make-up-productions.net/pages/posts /the-book-abecedarium-bestiarium---portraits-of-affinities-in-animal -metaphors-is-out-236.php.

REFERENCES

Deleuze, Gilles, and Félix Guattari. 1994. *What Is Philosophy?* Translated by Hugh Tomlinson and Graham Burchell. New York: Columbia University Press.

Leopold, Aldo. (1947) 1949. "On a Monument to the Pigeon." In *A Sand County Almanac, and Sketches Here and There*. New York: Oxford University Press.

CONTRIBUTORS

Michelle Bastian is a Chancellor's Fellow in the Edinburgh School for Architecture and Landscape Architecture at the University of Edinburgh. Her work focuses on the role of time in social processes of inclusion and exclusion. She has led UK AHRC-funded projects looking at issues to do with time, sustainability, and more-than-human communities. Her recent work has appeared in *Archival Science, New Formations*, and *Time and Society*. She is the lead editor of the collection *Participatory Research in More-than-Human Worlds* (2017). Find out more about her work at www.michellebastian.net.

Matthew Chrulew is ARC DECRA Research Fellow and leader of the Posthumanism-Animality-Technology research program in the Centre for Culture and Technology at Curtin University, Australia. With Dinesh Wadiwel, he edited *Foucault and Animals* (2016). His research interests include the history and philosophy of zoo biology, ethology, and conservation biology.

Vinciane Despret is Maître de conferences at the University of Liège and at the Free University of Brussels, Belgium. She was scientific curator of the exhibition "Bêtes et Hommes," held at the Grande Halle de la Villette, Paris, from September 12, 2007, to January 20, 2008, and has collaborated with philosophers, artists, choreographers, filmmakers, and scientists. She is the author or co-author of numerous books. Her books in English include *Our Emotional Makeup: Ethnopsychology and Selfhood* (2004), *Women Who Make a Fuss: The Unfaithful Daughters of Virginia Woolf* (with Isabelle Stengers; 2014), and *What Would Animals Say If We Asked the Right Questions?* (2016).

Rick De Vos is an adjunct research fellow in the Centre for Culture and Technology at Curtin University, Australia. His research focuses on species extinction and the way it functions as social and cultural practice. He has published essays in *Animal Studies Journal*, and in the edited collections *Knowing Animals* (2007) and *Animal Death* (2013). He is currently preparing a monograph examining the cultural significance of extinction and the way it operates discursively in dialogue with other aspects of cultural life.

James Hatley is a professor in the Department of Environmental Studies at Salisbury University, Maryland. He is a former member of the executive board of the International Association for Environmental Philosophy and is the author of *Suffering Witness: The Quandary of Responsibility After the Irreparable* (2000).

Deborah Bird Rose, FASSA, is an adjunct professor of environmental humanities at the University of New South Wales, Australia, and a founding co-editor of the journal *Environmental Humanities*. She is the author of numerous acclaimed books, including *Reports from a Wild Country: Ethics for Decolonisation* (2004), the recently re-released ethnography *Dingo Makes Us Human: Life and Land in an Australian Aboriginal Culture* (2009), and *Wild Dog Dreaming: Love and Extinction* (2011).

Thom van Dooren is associate professor of environmental humanities at the University of New South Wales, Australia, and a founding co-editor of the journal *Environmental Humanities*. His most recent book is *Flight Ways: Life and Loss at the Edge of Extinction* (2014).

Cary Wolfe is Bruce and Elizabeth Dunlevie Professor of English at Rice University, Houston, Texas, where he is also founding director of 3CT: Center for Critical and Cultural Theory. His recent books include *What Is Posthumanism?* (2010) and *Before the Law: Humans and Other Animals in a Biopolitical Frame* (2013). He is the founding editor of the Posthumanities book series of the University of Minnesota Press.

INDEX

Numbers in italics refer to pages on which illustration appear.

Abram, David, 30–31
abstract space, 112n.5
affirmation, 30, 36
'alalā (Hawaiian crow), 13, *186*,
    186–211; arrival of, in Hawai'i,
    189; captive breeding program for,
    189; diet and habits of, 188–189;
    extinction of, in the wild, 187, 189,
    196, 200–201; future
    reintroduction of, 194, 200;
    habitat of, 187–188; and Ka'u
    Forest Reserve, 194, 200, 201
    (*see also* Ka'u Forest Reserve); and
    Native Hawaiians, 191, 200–201;
    plants interdependent with,
    190–191, 206; vocalizations and
    learned behaviors of, diminished in
    captivity, 207
Andrea (student), 34, 38
animal culture: GLTs and learning,
    53, 69–70, 71–73; as missing
    element in conservation, 77n.14;

transmission of, lost in captivity,
    69, 71, 72–74, 207
Anthropocene, vii, 5; Hatley on, x, 28,
    31, 32; and jellyfish, 165; responses
    to, x, 7–8, 32; and time/
    temporality, xii–xiii. *See also*
    extinction; mass extinction
Arauz, Randell, 169–170, 171
Arfak astrapia (bird of paradise), 106.
    See also birds of paradise
Aristotle, 37
Atlantic Forest (Mata Atlântica;
    Brazil), 50–51. *See also* golden lion
    tamarin; Poço das Antas Biological
    Reserve
'*aumakua* (ancestral deity), 129–130,
    191
Australian Research Council, 141,
    208

Ball, Donna, 208
Ballou, Jonathan, 55, 76n.9

Banko, Paul, 208
Barnaby, Margaret, *186*
Barrow, Mark, 28
Basin and Range (McPhee), xii, xiii, xiv–xv
Bastian, Michelle, xi–xii, 12, 149–176. *See also* leatherback turtle
Beardsworth, Richard, xii
Beast and the Sovereign, The (Derrida), ix. *See also* Derrida, Jacques
Beck, Benjamin, 56, 59, 60–61, 64–65, 69, 70, 75n.6. *See also* golden lion tamarin
Beehler, Bruce, 101, 103, 104, 105
birds, xi, 103–104. *See also* 'alalā; birds of paradise; Passenger Pigeon; Pied Butcherbird
Birds of New Guinea and the adjacent Papuan Islands, The (Gould), 97–98
birds of paradise, 89–112; and colonialism, 89–92, 108–110; conservation efforts for, 102–103; Darwin on, 98–99; deaths and declining numbers of, 9, 100–101, 102–103; demand for specimens of, 96–97, 102; early descriptions and collection of, 89–92, 96–99; European paintings of, *88*, 97–98, 100; extinction of, 107, 110; females, 98; habits and habitat of, 103–104, 110; hybridity of, 104–107, 110, 111n.3, 111n.4; and Papuan people, 95–97; plumage of, 91, 98–99, 111n.1; and plume trade, 92, 93, 103, 108–110; rarity of, 105; taxonomy and classification of, 91, 98, 105 (*see also* birds of paradise: hybridity of). *See also* Elliot's Bird of Paradise
Birth, Kevin K., 175n.1, 176n.9

Bjorndal, Karen, 156
Black Sicklebill Bird of Paradise (*karanc*), 96, 102, 106. *See also* birds of paradise
Blanchot, Maurice, 136
Bolten, Alan, 156
Brañas, Roland, 153
Buddha, 21–22
Buddhism, 21–23, 35–36, 38, 40–41, 42–43, 45n.1. *See also* Kūkai
Bush, George W., 125
Butler, Judith, 135

captivity: 'alalā in, 207; breeding programs in, 52–54, 61, 76n.9, 189; cultural transmission lost in, 69, 71, 72–74, 207; GLTs in, 51, 52–54, 69–70, 72–74, 76n.8
care and caring: Puig de la Bellacasa on, 170; and rescuing leatherback turtles, 169–170; van Dooren on, 151, 172, 206; zoos' imperative for, 60, 61, 62, 64, 75n.7 (*see also* zoos). *See also* conservation efforts; ethics; responsibility; volunteers
Ching, Patrick, *116*
Chrulew, Matthew, xiv, 11, 49–78. *See also* golden lion tamarin
Cincinnati Zoo, xi, 217
circadian rhythms, 151, 175n.1
Clark, David, 136
Clifford, James, 197
clocks, 150, 151, 152, 159–160, 175n.1. *See also* leatherback turtle: satellite tracking of; time and temporality
coevolution, 190–191
Coimbra-Filho, Adelmar, 51, 70
colonization and colonialism: and birds of paradise and plume trade, 89–92, 96–97, 99–103, 108–110

(*see also* plume trade); colonial space, 108–110; and extinction, 12 (*see also* Elliot's Bird of Paradise); and habitat loss, 51; in Hawai'i, 198–200, 209n.5; in New Guinea, 92–95, 108–110, 111n.2 (*see also* New Guinea); and ornithology's attention to exotic/unknown, 96–98; and specimen collecting, 28

commercial fishing, impact of, 9, 120, 141n.2, 153–154, 168, 169–170. *See also* fishermen

community: communities involved with endangered animals, 119 (*see also* conservation efforts; volunteers; *specific species*); and ethics, 133–137; multispecies, 132–133, 135–137, 140–141; rational, 131–132, 142n.11; and time, 172 (*see also* time and temporality); unavowable, 136–137

Community of Those Who Have Nothing in Common, The (Lingis), 133–134

connection: and clocks, 150 (*see also* clocks); extinction as unraveling of, 193; interconnection of species and idea, 31; with leatherback turtles, 149–150, 169–170 (*see also* leatherback turtle)

conservation efforts, 2; in Atlantic Forest, 51; clarifying values and goals of, 209n.4; funding of, 171; and issues of colonization and occupation, 198–200, 205–206; narratives of, 55–56; and plume trade, 102–103; as "work of inheritance," 188. *See also* wildlife refuges and protected areas; *specific species*

contextual frameworks, 4

Costa Rica, 152–153, 156, 160–162

crow. *See* 'alalā

Crysdale, Shalan, 195–196, 200, 208

Culliney, Susan Moana, 190, 191

culture, of animals. *See* animal culture

Darwin, Charles, 98–99, 202–203. *See also* evolution, theory of

Dean, Warren, 50–51

death: as central to extinction processes, 8–9, 10; Clark on, 136; Derrida on, viii–ix; Levinas on responsibility and, 134–135; as necessity, 10, 73; risk of, by reintroduction, 75n.7 (*see also* golden lion tamarin). *See also* extinction; *specific species*

Dedication to PTT ID 56280 (Foster), 148

de-extinction, xi

deforestation, 50–51. *See also* forests

Derrida, Jacques, viii–x, xiii, 112n.5, 142n.12, 202, 210n.12, 210n.14

Despret, Vinciane, viii, xi, 13, 61, 217–221

De Vos, Rick, xi, 55, 89–112. *See also* birds of paradise; Elliot's Bird of Paradise

Diamond, Jared, 104

Diprose, Rosalyn, 211n.14

diversity, loss of, 196

dreaming, 30, 36

Durrell, Gerald, 56

Dutch East India Company, 93–94

East India Company (British), 94

ecological anachronisms, 191

Elliot, Daniel Giraud, 97

Elliot's Bird of Paradise, 12, *88*, 91, 106, 110, 111n.1. *See also* birds of paradise endangered animals: communities involved with, 119 (*see also* conservation efforts; volunteers); as predators of other endangered species, 160–162. *See also* extinction; *specific species*

Endangered Species Act, 128, 190

ethics: and community, 133–137; ethical responses to extinction, 7–10, 149; ethical work of inheritance, 203–206; responsible (ethical) inheritance, 203–206, 210n.14, 211nn.15–16

evolution, theory of, 28, 37–38, 202–203, 210n.11. *See also* inheritance

extinction: as biocultural phenomenon, 5; biocultural responses to, 2–6, 12 (*see also* extinction studies); choosing which species to remember, 32–33; and colonization, 12 (*see also* birds of paradise; colonization and colonialism); and community and human responsibility, 136–137; and death, 8–9, 10; determining the moment of, xi; ethical responses to, 7–10, 149; and generations, 9; Hatley on, 25–26; of Hawaiian bird species, 189–190; and humans, xiii–xiv; of humans, 37–38; and hunting (*see* hunting); hybrid loss vs. species extinction, 107; impact of, on world, 219–221; mass, xiii–xiv, 1–2, 26; McPhee on, xiv–xv; as multi-contextual phenomenon, viii; natural or unnatural, vii; as norm, 26; philosophical impact of, viii–ix, 27 (*see also* Hatley, James); remembrance of extinct species,

43–44; and specimen collection, 27–28; as unraveling, 193. *See also* conservation efforts; endangered animals; extinction studies; *specific species*

extinction studies, 2–6, 8–13. *See also* extinction

Extinction Studies Working Group, 2, 141

Ezo wolf, 29

fences. *See* Ka'u Forest Reserve

fishermen, 128, 130, 153–154, 169–170, 171–172. *See also* commercial fishing, impact of

flight way, definition of, 193

forests: Atlantic Forest and reserves, 50–51, 52, 54–55, 67, 72; deforestation, 50–51; in Hawai'i, 187–189, 190–191, 194–197, 201, 210n.9 (*see also* Ka'u Forest Reserve); and silviculture in Japan, 39–40

For What Tomorrow . . . A Dialogue (Derrida), 202

Foster, Kate, *148*

Foucault, Michel, 75n.5

fox sickness (Japan), 43

Frith, Clifford, 101, 105

frugivory, 103–104

Fuller, Errol, 106, 111n.4

"ghosts" of extinct or endangered animals: leatherback turtles, 156, 157; wolves, 25, 29–30, 41, 43. *See also* 'alalā

Gilbert, Scott, 193

Gilliard, E. Thomas, 102, 111n.3

golden lion tamarin (GLT; *mico*), *48*, 49–78; breeding program for, 52–54, 61, 76n.9; in captivity, 51,

52–54, 69–70, 72–74, 76n.8; deaths of, reintroduced into the wild, 54–55, 58–62, 66, 68–70, 75nn.6–7, 76n.8, 78n.16; declining numbers of, in the wild, 49, 51–52, 57; description, taxonomy, and habitat of, 50, 74n.1; efforts to conserve, in the wild, 11, 52–54; future of, 73–74; and habitat loss, 49, 50–51, 57; hazards faced by, in the wild, 49–50, 59–61, 63–64, 75n.6; learned skills of, 53, 69–70, 71–73; lessons of reintroduction experiment, 70–73, 78n.16; plasticity of, 67, 69–70, 71–72, 74; popularity of, 56; post-release management (soft release) of, 65–68, 73, 77nn.9–10; price/costs of conservation efforts for, xiv, 11, 57–58, 61–62, 70, 74n.2, 75nn.5–6; rehabilitation (pre-release enrichment) of, 62–65, 70; reintroduction of, as short-term solution, 74n.3; reintroduction of, into the wild, 11, 49–50, 54–55, 58–62; reintroduction program for, as experiment, 11, 68–71, 78n.16; reintroduction program for, deemed success, 55–58, 66; reintroduction program for, wider impact of, 66–68

Gould, John, 91, 97

Gray, Russell, 192

Great Britain, 102–103

green turtle, 161, 162

grief and mourning, 13, 37, 219–221

Griffiths, Paul, 192

Guji, Asahi, 39–40

habitat loss/decline: deaths related to, 9; and GLTs, 49, 50–51, 57; and

Honshu wolf, 39, 42; and jaguars, 160; and species interdependence, 104. *See also* wildlife refuges and protected areas; *specific species*

haiku, 45nn.2–4; on burning prayer sticks, 40; on camellias, 24; on carp in a pond, 35; on deserted teahouse, 24; on Honshu wolf, 19; on stones as boats of prayer, 22; on trees, 23; on winter mountains and wolves (Shiki), 29; on wolf scat and cold (Issa), 25

halapepe (Hawaiian plant), 190, 191

Handelman, Susan, 134, 140

Haraway, Donna, 3–4, 13n.4, 14n.7, 193

Hart, Kevin, 142n.12

Hatley, James, viii–x, 19–46; on ethical demands of writing, 13n.5; on extinction as disappearance and question, 11; poetry and haiku by, 23, 24, 35, 40, 44–45, 45n.2; on transhuman etiquette, viii, 27, 30. *See also* Honshu wolf

Hawaiian crow. *See* ʻalalā

Hawaiian Islands, 201, 208n.2; bird species in, 189–190; conflict over land ownership/use in, 195–200, 208n.2, 209n.5, 210n.9; federally protected areas (NWHI) in, 125–128; forests in, 187–189, 190–191, 194–197, 201, 210n.9 (*see also* Kaʻu Forest Reserve); and Hawaiian monk seals, 118, 124–125, 128–131; human settlement of, 120–121, 129–130, 194; introduced ungulates in, 188, 189, 190; loss of ʻalalā felt in, 191–192; Native culture in, 129–131, 191–192, 195–198, 201; pigs and pig hunting in, 188,

Hawaiian Islands (*continued*)
195–198, 201, 208n.2; U.S.
occupation and colonization of,
198–200, 209n.5. *See also* 'alalā;
Hawaiian monk seal; Kaua'i; *specific
islands and individuals*

Hawaiian monk seal, 12, *116*, 117–142;
conservation efforts for, 121–128;
deaths and commercial hunting of,
9, 120, 126–127, 128–129, 131,
141n.2; habits, diet, and habitat of,
118, 120; haul outs (sleeping on
beaches) by, 118–119, 122–123;
Hawaiians divided about, 128–131;
and human settlement of Hawai'i,
120–121, 129; K P2 (Ho'ailona),
117–118, 119, 124, 128–129, 132;
popularity of, 118, 123–124,
128–129; population decline of,
120, 121, 126–127; prayers for,
139–140; in protected vs.
unprotected areas, 125–128; pups
and mothers, 123–124, 126–127,
138–139; risks of human
interaction with, 120, 124, 128;
susceptibility of, to disease, 128,
142n.10; taxonomy and
classification of, 119–120; and
volunteers, 118, 121–125, 131–133,
137–139

Hawaiian Monk Seal (*Monachus
schauinslandi*) (Ching), *116*

Hawaiian Monk Seal Research
Program, 126–127

Hawaiian TurtleWatch (mapping
tool), 154, 175n.5

Hegel, Georg Wilhelm Friedrich,
xvn.2

Heidegger, Martin, ix–x

Heraclitus, 35–36, 37

Ho'ailona. *See* K P2

ho'awa (Hawaiian tree), 190, 191, 206

Hogan, Linda, 33

Hongu (Japan), 34–35

Honshu (Japan), 20–21, 23–24,
39–40. *See also* Honshu wolf;
Kumano Kodō

Honshu wolf (Ōkami), 11, *18*, 19–46;
beliefs and folktales about, 31,
36–37, 42; deaths of, 9; ecological
role of, 42; extinction of, 20,
25–27, 31–34, 42; "ghosts" of, 25,
41, 43; and habitat loss/decline, 39,
42; lessons offered by, 41, 43;
names and naming of, 20, 31,
32–34; response to disappearance
of, viii, 26–27, 36–37, 43–45;
specimen (last wolf shot) of, 9,
27–28. *See also* Hatley, James

Houghton, Jonathan, 163

humans: arrival of, in Hawaiian
Islands, 120–121, 129–130, 194;
communities of, involved with
endangered animals, 119 (*see also*
community; conservation efforts;
Hawaiian monk seal; volunteers;
zoos); delusion of control by, 41;
and extinction events, 1, 6,
136–137 (*see also* extinction; *specific
extinct and endangered species*);
extinction of, 37–38; and
interspecies dependence, 193–194;
and jellyfish, 163–164; nonhuman
animals living in proximity to,
118–119; and time, xii–xiii (*see also*
time and temporality); Žižek on,
xvn.2

Hume, Julian, 91, 111n.1, 112n.4

hunting: of birds of paradise in New
Guinea, 12, 96, 100–101 (*see also*
plume trade); hunters and
conservation, 195–198, 199; of pigs

in Hawai'i, 188, 195–198, 201, 208n.2
hybridity, 12, 104–107, 110, 111n.3, 111n.4, 112n.5

inheritance, 13; biocultural, 192–194, 196, 203, 210n.11, 210n.13; cultural, lost in captivity, 69, 71, 72–74, 207; Derrida on, 202–203, 210n.12, 210n.14; extinction as termination of, 9; histories as acts of, 202; responsible (ethical), 203–206, 210n.14, 211nn.15–16. *See also* evolution, theory of
International Union for Conservation of Nature, 52, 104, 121
interspecies dependence: birds and trees, 103–104, 189, 190–191; erasure of, in depictions of birds of paradise, 97; and human nature, 194; jaguars and their prey, 160–162, 166; leatherback turtles and jellyfish, 162–165, 167; recognition of, in Hawaiian culture, 129. *See also* golden lion tamarin
Iredale, Tom, 111n.3
Issa, Kobayashi, 25

jack fish (*ulua*), 127–128
jaguar, 160–162, 166
James, Michael, 171–172
Japan. *See* Honshu; Honshu wolf; Kumano Kodō
Jasu (teacher, volunteer ranger), 40
jellyfish, 162–165, 167, 176n.11
Jersey Wildlife Preservation Trust (Durrell Wildlife Conservation Trust; Great Britain), 52, 56
Jiso (docent), 21–23

Ka'awa, Nohea, 195–196, 208
Kaneholani, Daniel, 128
Kato, Kumi, 20–21, 39, 43
Kaua'i: conservation volunteers on, 121–125, 137–139; Hawaiian monk seals on, 128, 137–140; Po'ipū Beach on, 124–125, 131, 142n.6. *See also* Hawaiian Islands; Hawaiian monk seal
Kaua'i Monk Seal Conservation Hui, 122. *See also* Hawaiian monk seal: and volunteers
Kaua'i Monk Seal Watch Program, 122. *See also* Hawaiian monk seal: and volunteers; Robinson, Tim
Kauanui, J. Kehaulani, 199, 209n.5
Ka'u Forest Reserve (Hawai'i), 194–197, 198–200, 201, 210n.9
Kauka, Kumu Sabra, 131
Kemarre Turner, Margaret, 30
Kierulff, Maria Celia M., 78n.16
Kirksey, Eben, 111n.2
Kirsch, Stuart, 100
Kitayama Moon: Toyohara Sumiaki (Yoshitoshi), *18*
Kittinger, John, 121, 129–130
Kleiman, Devra, 53, 56, 60, 61. *See also* golden lion tamarin
Knight, John, 44
Kolbert, Elizabeth, 6
K P2 (Ho'ailona; captive Hawaiian monk seal), 117–118, 119, 124, 128–129, 132
Kūkai, 20, 22–23, 45n.1
Kumano Kodō, 20–25, 29, 35, 39–40, 43. *See also* Honshu; Honshu wolf
Kumulipo (Hawaiian origin story), 129, 201

leatherback turtle, 12, *148*, 149–176; of Atlantic Ocean, 161–165; building connections with, 149–150; coastal, 155–156; commercial fishing as threat to, 153–154, 168, 169–170; conservation efforts for, 153, 154–155, 168, 171–172, 175n.5; of eastern Pacific Ocean, 153–157; hatchlings, 175n.3; and jellyfish, 162–165, 167; migration of, 155–157; nesting (egg laying) by, 152–153, 156, 160, 173; plastic eaten by, 166–167; predators of, 160–162; PTT ID 56280, *148*, 155–157; rescues of, 169; satellite tracking of, 154–155, 175n.6; speed of, xii; and synchrony, 153–154, 156, 161–164, 175n.3; as time givers, 151; in unexpected places, 173–174
Lefebvre, Henri, 112n.5
Leopold, Aldo, 219
Lesson, René Primevère, 89–90, 92
Leunig, Michael, 139
Levinas, Emmanuel, 133, 134–135, 139
Lieberman, Alan, 208
Lili'uokalani (queen of Hawai'i), 198
Lingis, Alphonso, 131–132, 133–134, 139
Littnan, Charles, 126–128, 132
Lost Birds of Paradise, The (Fuller), 106
Louise (artist, teacher), 40
loulu palms, *186*, 190, 191

Malay Archipelago, The (Wallace), 90–91. *See also* Wallace, Alfred Russel
Male Passenger Pigeon (*Ectopistes migratorius*) (Hayashi and Toda), *216*
Mallinson, Jeremy, 55–56

Marine Mammal Physiology Project, 117. *See also* Williams, Terrie
Martin, Kathleen, 171–172, 174
mass extinction, xiii–xiv, 1–2, 26. *See also* extinction; *specific species*
Mata Atlântica. *See* Atlantic Forest
Mathews, Freya, 30–31
Mayr, Ernst, 111n.3
McPhee, John, xii, xiii, xiv–xv
metapopulation management, 67. *See also* reintroduction
Meyer, Adolf Bernard, 91, 92, 111n.1
MHI (main Hawaiian Islands). *See* Hawaiian Islands
Mihashi, Toshio, 19
Miyaji, Tempei, 22
monk seals, 119–120. *See also* Hawaiian monk seal
Monograph of the Paradiseidæ, or Birds of Paradise, A (Elliot), 97–98
Monograph of the Paradiseidæ, or Birds of Paradise, and Ptilonorhynchidæ, or Bower-Birds (Sharpe), *88*, 97–98
Mooallem, Jon, 135–136
Mooney, Edward, 36
mourning. *See* grief and mourning
multiple universes, 38
multispecies studies, 4–5
Munro, George, 187

Naas, Michael, ix
Nachi Falls and Shinto temple (Honshu), 21, 39
National Oceanic and Atmospheric Administration (NOAA), 121, 130, 132
National Zoo (Washington, D.C.), 52, 54, 56
Natural History of Monkeys, The (Jardine), *48*
nature, as concept, vii

Nature Conservancy, 195

Nealon, Jeffrey, 75n.5

New Guinea: ethnic and linguistic groups in, 93, 94; European colonization and representation of, 92–95, 108–110, 111n.2; geography and ecology of, 93, 94, 103–104; scientific exploration of, 89–92. *See also* birds of paradise; Elliot's Bird of Paradise; Papuan people

NOAA. *See* National Oceanic and Atmospheric Administration

nonhuman life, ix–x. *See also specific species*

nonscientific knowledges, 4

Northwestern Hawaiian Islands (NWHI), 125–128. *See also* Hawaiian monk seal

Nova Scotia Leatherback Turtle Working Group (NSLTWG), 171–172

NWHI. *See* Northwestern Hawaiian Islands

Obama, Barack, 125

oil, xiii

Ōkami. *See* Honshu wolf

One Hundred Aspects of the Moon (*Tsuki hyakushi*) (Yoshitoshi), *18*

ornithology, 97–99, 101–102, 105–107. *See also* birds of paradise; specimens

Ortega y Gasset, José, 36

Orthogenetic Evolution in the Pigeons (Whitman), *216*

Ortiz de Retes, Yñigo, 94–95

Papahānaumokuākea Marine National Monument (Hawai'i), 125–127

Papua New Guinea, 93–94. *See also* birds of paradise; New Guinea

Papuan people, 93–97, 99, 100–101, 108–109, 111n.2. *See also* New Guinea

Passenger Pigeon, viii, xi, 9, *216*, 217–221

peccary, 160

philosophical discourse: affirmation and amplification, 30–31, 35–36; Japanese philosophy, 39, 40 (*see also* Buddhism); on naming of extinct species, 32–34; perpetuation of, 36; on philosophical impact of extinction, viii–ix, 27. *See also* community; ethics; extinction; inheritance; time and temporality

Pied Butcherbird, 159

pigs and pig hunting, in Hawai'i, 188, 195–198, 201, 208n.2

pilgrimage. *See* Kumano Kodō

plasticity: of animals, 77n.15; of GLT, 67, 69–70, 71–72, 74 (*see also* golden lion tamarin)

plastics, 166–167

Playa Grande (Costa Rica), 152–513, 156. *See also* leatherback turtle

plume trade, 92, 93, 103, 108–110. *See also* birds of paradise

Poço das Antas Biological Reserve (Brazil), 52, 52, 54–55, 67, 72. *See also* golden lion tamarin

Po'ipū Beach (Kaua'i), 124–125, 131, 142n.6

polygynous birds, 103–104. *See also* birds of paradise

prayers, physical, 139–140

primates. *See* golden lion tamarin; humans

Psalm 19, 19

PTT ID 56280 (leatherback turtle), *148*, 155–157
Puerto Rico, 156
Puig de la Bellacasa, Maria, 170

rational communities, 131–132, 142n.11
recuperative work, 205–206
reflected images, 35–36
Reichenow, Anton, 105
reintroduction: animal deaths risked by, 75n.7; and animal plasticity, 78n.15; and captive breeding programs, 61–62, 76n.9; as short-term solution, 74n.3; zoo biological techniques applied in the wild, 65–68; and zoos and zookeepers, 60–61. *See also* golden lion tamarin
Replogle, John, 195–196, 208
responsibility: responding ethically to extinction, 7–10; responsible (ethical) inheritance, 203–206, 210n.14, 211nn.15–16. *See also* community; ethics
Ritte, Walter, 130–131, 135–136
Robinson, Tim, 122, 138–140
Roosevelt, Theodore, 125
Rose, Deborah Bird, 12, 117–142; on "adding flesh" to abstract analysis of time, 176n.8; on recuperative work, 205–206; on sequence and synchrony in multispecies temporality, 150–151, 193; on shared lifeworlds, 150; on survivors of "extinct" species, xi; on temporality in a multispecies world, xi–xii; on writing ethically about extinction, 149. *See also* Hawaiian monk seal

San Diego Zoo (California), 189
Sanguine Moon (Barnaby), *186*
satellite tracking, of leatherback turtles, 154–155, 175n.6
"Saving the Lion Marmoset" (conference), 52
seed dispersal, 104, 189, 190–191
sharks, 127–128
Sharma, Sarah, 170, 172
Sharpe, Richard Bowdler, 97–98, 101–102
Shiki, Masoaka, 29
Shillinger, George, 154–156, 170
shotguns, 101
Sixth Extinction, The (Kolbert), 6
Smith, Mick, 14n.7
sound and silence, 22–23
space: abstract, 112n.6; colonial, 108–110 (*see also* colonization and colonialism); Lefebvre on, 112n.5
Spain, 93–95
species interdependence. *See* interspecies dependence
species loss. *See* endangered animals; extinction; mass extinction; *specific species*
specimens: collection, preservation, and study of, 27–28, 43, 90; private market for, 96–97, 98; and taxonomy and classification, 91, 105–107. *See also specific species*
Spotila, James, 153
Springer, Hannah Kihalani, 200–201, 204–205, 208
Star, Susan Leigh, 6
Stengers, Isabelle, 68
Steutermann, Kim, 124–125, 128, 133, 138
Stoekl, Alan, xiii
storytelling, 3–4, 13n.5, 30, 55–56, 196, 202, 207

Streseman, Erwin, 91, 105–106, 110, 111nn.3–4
Switzer, Rich, 208

Temminck, Coenraad, 31
temporality. *See* time and temporality
Thirteen Gold Monkeys (Beck), 56, 59, 60–61. *See also* Beck, Benjamin; golden lion tamarin
time and temporality: and Bastian's work, 12, 150–152, 157–160, 168–169, 173–175 (*see also* Bastian, Michelle; leatherback turtle); clocks and connection, 150, 151, 152, 159–160, 170; differential speeds of, in multispecies world, xii–xiii, 9–10; and extinction, 8; and extinction studies, 9–10; and jellyfish, 164–165; multispecies knots of, xii, xv, 150–151, 154 (*see also* Bastian, Michelle); sequence and synchronies of, 150–151, 161–162, 166 (*see also* leatherback turtle); as tethered and collective, 170–171; and "triangulation" as term, 176n.9
Tortuguero (Costa Rica), 160–162
tradition, formation and transmission of, 197, 202
transhuman etiquette, viii, 27, 30
trees, fruiting, 103–104. *See also* seed dispersal
Tsing, Anna, 4
Tsurugi-no-yama (Sword Mountain; Honshu), 22–23

unavowable community (death-age community), 136–137
União Biological Reserve (Brazil), 55
unwork, 137

van Dooren, Thom, 186–211; on albatrosses, 141n.3; foundational work of, xi–xii; on inheritance, 12, 13 (*see also* inheritance); on questions of care, 172 (*see also* care and caring). *See also* 'alalā
volunteers: and Hawaiian monk seals, 121–125, 131–133, 137–139; and leatherback turtles, 169–170, 171–172; motivation and justification of, 133

Waikiki Aquarium (Hawai'i), 117–118, 119, 124, 132
Wallace, Alfred Russel, 90–91, 92, 107–108
Walters, Michael, 91, 111n.1, 112n.4
Ward, Edwin, 91, 92
Weisman, Alan, 166
Weston, Anthony, 27
West Papua (Indonesia), 93–94, 111n.2. *See also* New Guinea
whales, xii, 43, 171
wild animals, encounters with, 138. *See also specific species*
wildlife refuges and protected areas: Atlantic Forest (Brazil), 51, 52, 54–55, 67, 72 (*see also* golden lion tamarin); Ka'u Forest Reserve (Hawai'i), 194–197, 198–200, 201 (*see also* 'alalā); Northwestern Hawaiian Islands, 125–127 (*see also* Hawaiian monk seal); Papahānaumokuākea Marine National Monument (Hawai'i), 125–127; Playa Grande (Costa Rica), 152–153, 156 (*see also* leatherback turtle)
Williams, Terrie, 117, 124, 126, 142n.10
Wolf, Joseph, 97

Wolfe, Cary, vii–xvi, 7
wolves, 44. *See also* Honshu wolf
writing, hybridity as, 106–107,
112n.5
Wyschogrod, Edith, 133, 136–137

Yoshitoshi, Tsukioka, *18*
Yunomine temple and Buddha
(Honshu), 40–41

Žižek, Slavoj, xvn.2
zoos, 65–68; and ʻalalā, 189; and
captive/wild distinction, 60, 70,
77n.10; deaths of animals in, 75n.7;
enrichment for animals in, 63–64,
70; and GLTs, 11, 51, 52–57, 60–70
(*see also* golden lion tamarin); last
Passenger Pigeons in, xi, 217; and
reintroduction into the wild, 60

GPSR Authorized Representative: Easy Access System Europe, Mustamäe tee
50, 10621 Tallinn, Estonia, gpsr.requests@easproject.com